my **revisi⏻n** notes

WJEC GCSE
MATHEMATICS
HIGHER

Gareth Cole
Karen Hughes
Joe Petran
Keith Pledger
Linda Mason

HODDER
EDUCATION
AN HACHETTE UK COMPANY

The Publishers would like to thank the following for permission to reproduce copyright material.

Acknowledgements

Every effort has been made to trace all copyright holders, but if any have been inadvertently overlooked, the Publishers will be pleased to make the necessary arrangements at the first opportunity.

Although every effort has been made to ensure that website addresses are correct at time of going to press, Hodder Education cannot be held responsible for the content of any website mentioned in this book. It is sometimes possible to find a relocated web page by typing in the address of the home page for a website in the URL window of your browser.

Hachette UK's policy is to use papers that are natural, renewable and recyclable products and made from wood grown in sustainable forests. The logging and manufacturing processes are expected to conform to the environmental regulations of the country of origin.

Orders

Bookpoint Ltd, 130 Park Drive, Milton Park, Abingdon, Oxon OX14 4SE.
Telephone: (44) 01235 827720.
Fax: (44) 01235 400454.
Email education@bookpoint.co.uk
Lines are open from 9 a.m. to 5 p.m., Monday to Saturday, with a 24-hour message answering service. You can also order through our website: www.hoddereducation.co.uk

ISBN: 978 1 4718 8253 1

© Keith Pledger, Gareth Cole, Joe Petran, Karen Hughes, Linda Mason 2017

First published in 2017 by

Hodder Education,
An Hachette UK Company
Carmelite House
50 Victoria Embankment
London EC4Y 0DZ

www.hoddereducation.co.uk

Impression number   10 9 8 7 6 5 4 3 2 1

Year   2021 2020 2019 2018 2017

Cover photo © koya79/Thinkstock/iStockphoto/Getty Images
Typeset in Integra Software Services Pvt. Ltd., Pondicherry, India
Printed in Spain

A catalogue record for this title is available from the British Library.

Get the most from this book

Welcome to your Revision Guide for the WJEC GCSE Mathematics Higher course. This book will provide you with sound summaries of the knowledge and skills you will be expected to demonstrate in the exam, with additional hints and techniques on every page. Throughout the book, you will also find a wealth of additional support to ensure that you feel confident and fully prepared for your GCSE Maths Higher examination.

This Revision Guide is divided into four main sections, with additional support at the back of the book. The four main sections cover the four mathematical themes that will be covered in your course and examined: Number, Algebra, Geometry & Measures and Statistics & Probability.

Features to help you succeed

Each theme is broken down into one-page topics as shown in this example:

The knowledge you have learned on your course is reduced to the key rules for this topic area. You will need to understand and remember these for your exam.

Worked examples are provided to remind you how the rules work. Each rule is highlighted next to where it is being used.

Exam-style questions provide real practice on the topic area, with allocated marks so you can see the level of response that is required.

Each page is given a level of difficulty so you can understand the level of challenge. Broadly speaking, Low denotes grades 1-2, Medium = 3, High = 4-5.

Areas where common errors are often made are highlighted to help you avoid making similar mistakes.

These are the terms and phrases you will need to remember for this topic.

Hints and techniques will suggest what to remember or how to approach an exam question.

Each theme section also includes the following:

Pre-revision check

Each section begins with a test of questions covering each topic within that theme. This is a helpful place to start to see if there are any areas which you may need to pay particular attention to in your revision. To make it easier, we have included the page reference for each topic page next to the question.

Exam-style question tests

There are two sets of tests made up of practice exam-style questions to help you check your progress as you go along. These can be found midway through and at the end of each theme. You will find **Answers** to these tests at the back of the book.

At the end of the book, you will find some very useful information provided by our assessment experts:

The language used in mathematics examinations

This page explains the wording that will be used in the exam to help you understand what is being asked of you. There are also some extra hints to remind you how to best present your answers.

Exam technique and formulae that will be given

A list of helpful advice for both before and during the exams, and confirmation of the formulae that will be provided for you in the exams.

Common areas where students make mistakes

These pages will help you to understand and avoid the common misconceptions that students sitting past exams have made, ensuring that you don't lose important marks.

One week to go....

Reminders and formulae for you to remember in the final days before the exam.

Tick to track your progress

Use the revision planner on pages v to viii to plan your revision, topic by topic. Tick each box when you have:
- worked through the pre-revision check
- revised the topic
- checked your answers

You can also keep track of your revision by ticking off each topic heading in the book. You may find it helpful to add your own notes as you work through each topic.

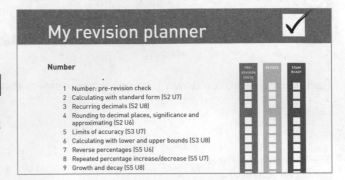

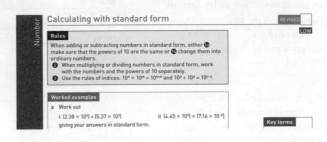

My revision planner

Number

		PRE-REVISION CHECK	REVISED	EXAM READY
1	Number: pre-revision check			
2	Calculating with standard form			
3	Recurring decimals			
4	Rounding to decimal places, significance and approximating			
5	Limits of accuracy			
6	Calculating with lower and upper bounds			
7	Reverse percentages			
8	Repeated percentage increase/decrease			
9	Growth and decay			
10	Mixed exam-style questions			
11	Working with proportional quantities			
12	The constant of proportionality			
13	Working with inversely proportional quantities			
14	Formulating equations to solve proportion problems			
15	Index notation and rules of indices			
16	Fractional indices			
17	Surds			
18	Mixed exam-style questions			

Algebra

		PRE-REVISION CHECK	REVISED	EXAM READY
19	Algebra: pre-revision check			
21	Simplifying harder expressions and expanding two brackets			
22	Using complex formulae and changing the subject of a formula			
23	Identities			
24	Using indices in Algebra			
25	Manipulating more expressions; algebraic fractions and equations			
26	Rearranging more formulae			
27	Special sequences			
28	Quadratic sequences			

	PRE-REVISION CHECK	REVISED	EXAM READY

29 nth term of a quadratic sequence

30 The equation of a straight line

31 Plotting quadratic and cubic graph

32 Finding equations of straight lines

33 Polynomial and reciprocal functions

34 Perpendicular lines

35 Exponential functions

36 Trigonometric functions

38 Mixed exam-style questions

40 Trial and improvement

41 Linear inequalities

42 Solving simultaneous equations by elimination and substitution

43 Using graphs to solve simultaneous equations

44 Solving linear inequalities

46 Factorising quadratics of the form $x^2 + bx + c$

47 Solve equations by factorising

48 Factorising harder quadratics and simplifying algebraic fractions

49 The quadratic equation formula

50 Using chords and tangents

52 Translations and reflections of functions

53 Area under non-linear graphs

55 Mixed exam-style questions

Geometry and Measures

57 Geometry and Measures: pre-revision check

59 Working with compound units and dimensions of formulae

60 Congruent triangles and proof

61 Proof using similar and congruent triangles

62 Circle theorems

63 Pythagoras' theorem

64 Arcs and sectors

65 The cosine rule

		PRE-REVISION CHECK	REVISED	EXAM READY
67	The sine rule	☐	☐	☐
68	Loci	☐	☐	☐
69	Mixed exam-style questions	☐	☐	☐
71	Similarity	☐	☐	☐
72	Trigonometry	☐	☐	☐
73	Finding centres of rotation	☐	☐	☐
75	Enlargement with negative scale factors	☐	☐	☐
77	Trigonometry in 2D and 3D	☐	☐	☐
78	Volume and surface area of cuboids and prisms	☐	☐	☐
80	Enlargement in two and three dimensions	☐	☐	☐
81	Constructing plans and elevations	☐	☐	☐
82	Surface area and volume of 3D shapes	☐	☐	☐
83	Area and volume in similar shapes	☐	☐	☐
84	Mixed exam-style questions	☐	☐	☐

Statistics and Probability

		PRE-REVISION CHECK	REVISED	EXAM READY
86	Statistics and Probability: pre-revision check	☐	☐	☐
88	Using grouped frequency tables	☐	☐	☐
89	Inter-quartile range	☐	☐	☐
91	Displaying grouped data	☐	☐	☐
93	Histograms	☐	☐	☐
95	Mixed exam-style questions	☐	☐	☐
97	Working with stratified sample techniques and defining a random sample	☐	☐	☐
98	The multiplication rule	☐	☐	☐
100	The addition rule and Venn diagram notation	☐	☐	☐
102	Conditional probability	☐	☐	☐
103	Mixed exam-style questions	☐	☐	☐

Exam preparation

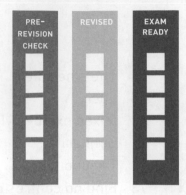

104 The language used in mathematics examinations

105 Exam technique and formulae that will be given

106 Common areas where students make mistakes

112 One week to go...

115 Answers

Number: pre-revision check

Check how well you know each topic by answering these questions. If you get a question wrong, go to the page number in brackets to revise that topic.

1 Work out:
  a $(1.5 \times 10^3) \div (2.8 \times 10^{-1})$
  b $(9.42 \times 10^2) + (1.36 \times 10^3)$ (Page 2)

2 a Write 1.037 373 737 … in recurring decimal notation.
  b Convert $0.\dot{1}\dot{8}$ to a fraction.   (Page 3)

3 A number is given as 8.37 correct to 2 decimal places.
  Write down the lower and upper bounds of this number.   (Page 5)

4 5.62 and 2.39 are written correct to 2 decimal places. Write down the lower and upper bounds of:
  a $5.62 - 2.39$
  b 5.62×2.39 (Page 6)

5 The average cost of a new house in Brinton has increased by 6%. The average cost is now £118 720. What was the average cost before this increase?   (Page 7)

6 Rehan buys a car for £12 000. The value of the car depreciates by 10.5% each year for the first 3 years. What is the value of Rehan's car after 3 years?   (Page 8)

7 Mirza invests £5000 into an account paying compound interest at a rate of 3.2% p.a.
  How many years will pass before this investment is first worth more than £6000?   (Page 9)

8 Which of these tables of values illustrate direct proportion and which illustrates inverse proportion?   (Page 12)

A

x	1	2	5	12	20
y	300	150	60	25	15

B

x	10	20	30	40	50
y	2.5	5	7.5	10	12.5

C

x	0.1	0.2	0.3	0.4	0.5
y	180	90	60	45	36

9 T is inversely proportional to x. When $x = 1.4$, $T = 25$. Write down a formula for T in terms of x.   (Page 13)

10 P is directly proportional to the square root of A. When $A = 25$, $P = 30$.
  Write down a formula for P in terms of A.   (Page 14)

11 Work out the value of the following.
  a $(2^4 \div 2^{-3}) \times 2^{-5}$
  b $(10^9 \times 10^{-4} \div 10^3)^2$ (Page 15)

12 Work out the value of the following.
  a $27^{\frac{2}{3}}$
  b $3125^{-\frac{1}{5}}$ (Page 16)

13 a Simplify $\sqrt{700}$.
  b Rationalise the denominator of $\frac{1}{2\sqrt{3}}$.   (Page 17)

Calculating with standard form

Rules

When adding or subtracting numbers in standard form, either **1a** make sure that the powers of 10 are the same or **1b** change them into ordinary numbers.

2 When multiplying or dividing numbers in standard form, work with the numbers and the powers of 10 separately.

3 Use the rules of indices: $10^n \times 10^m = 10^{n+m}$ and $10^p \div 10^q = 10^{p-q}$.

Worked examples

a Work out

i $(2.38 \times 10^5) + (5.37 \times 10^3)$

ii $(4.45 \times 10^3) \times (7.16 \times 10^{-2})$

giving your answers in standard form.

Answers

i $2.38 \times 10^5 + 5.37 \times 10^3$ **OR** $2.38 \times 10^5 + 5.37 \times 10^3$

1a $= 238 \times 10^3 + 5.37 \times 10^3$ **1b** $= 238000 + 5370$
$= (238 + 5.37) \times 10^3$ $= 243370$
$= 243.37 \times 10^3$ $= 2.4337 \times 10^5$
$= 2.4337 \times 10^5$

ii $4.45 \times 10^3 \times 7.16 \times 10^{-2}$

2 $= 4.45 \times 7.16 \times 10^3 \times 10^{-2}$
$= 31.862 \times 10^1$ **3**
$= 3.1862 \times 10 \times 10^1$
$= 3.1862 \times 10^2$

b How many times greater than 9.27×10^3 is 3.16×10^8?

Answer
Number of times is $(3.16 \times 10^8) \div (9.27 \times 10^3)$

$= 3.16 \div 9.27 \times 10^8 \div 10^3$ **2**
$= 0.34088... \times 10^5$ **3**
$= 3.4088... \times 10^{-1} \times 10^5$ **2**
$= 3.4088... \times 10^4$

Key terms

Standard form

Ordinary number

Powers, indices

Exam tip

Give your answer in standard form if the question asks for it.

Look out for

Be careful not to write:

$3.16 \div 9.27 \times 10^8 \times 10^3$, the powers in 10 also need to be divided.

Exam-style questions

1 $p = 6.32 \times 10^4$ $q = 7.15 \times 10^{-2}$
Work out
  a pq **[2]**
  b $(p + q)^2$ **[2]**
Give your answers in standard form.

2 The diameter of a water molecule is 2.9×10^{-8} cm.
One nanometre $= 1 \times 10^{-9}$ metres.
What is the diameter of this water molecule in nanometres?
Give your answers in standard form. **[2]**

3 One light year $= 9.461 \times 10^{12}$ km.
The average distance from the Sun to Earth $= 1.496 \times 10^8$ km
How many times greater is one light year than the average distance from the Sun to Earth?
Give your answers in standard form. **[2]**

Exam tip

Do not try to do the whole of each calculation on your calculator. Write down each stage of your working.

CHECKED ANSWERS

Recurring decimals

Rules

1. To interpret recurring decimal notation (dots above the 1st and last digit), repeatedly write down this range of numbers; for example, $1.2\dot{3}4\dot{5} = 1.2345345345345\ldots$
2. To write a recurring decimal as a fraction, multiply the decimal by suitable powers of 10 such that the difference between **two** subsequent answers is a rational number.

Worked examples

a Convert $1.\dot{7}$ to a fraction.

Answer

Let $x = 1.\dot{7} = 1.77777\ldots$ **①**

$10x = 1.77777\ldots \times 10 = 17.77777\ldots$ **②**

$10x - x = 17.77777\ldots - 1.77777\ldots = 16$ (a rational number)

$9x = 16$, so $x = 1.\dot{7} = \frac{16}{9}$

b Convert $2.8\dot{4}\dot{5}$ to a fraction.

Answer

Let $x = 2.8\dot{4}\dot{5} = 2.8454545\ldots$ **①**

$10x = 2.8454545\ldots \times 10 = 28.454545\ldots$ **②**

$1000x = 2.8454545\ldots \times 1000 = 2845.454545\ldots$

$1000x - 10x = 2845.454545\ldots - 28.454545\ldots = 2817$ (a rational number)

$990x = 2817$, so $x = 2.8\dot{4}\dot{5} = \frac{2817}{990}$

Key term

Recurring decimal

Exam tips

Never try to use your calculator to convert; the question will always imply an algebraic method.

If the question is on a calculator paper, check your answer.

Exam-style questions

1 Prove algebraically that the recurring decimal $0.5\dot{4}$ can be written as the fraction $\frac{49}{90}$. **[3]**

2 Prove algebraically that the recurring decimal $0.\dot{4}2\dot{5}$ can be written as the fraction $\frac{425}{999}$. **[3]**

3 Prove algebraically that the difference between $2.1\dot{8}$ and $1.\dot{1}$ can be written as the fraction $1\frac{7}{99}$. **[3]**

CHECKED ANSWERS

Exam tip

A common error is to show that the fraction $\left(\frac{49}{90}\right)$ can be converted, by division, into the given recurring decimal. This would get no marks since an algebraic approach is required.

Rounding to decimal places, significance and approximating

Rules

❶ To round a number to decimal places, look at the next number after the required number of decimal places; ❶ⓐ if it is 5 or above, increase the previous place number by 1; ❶ⓑ if it is less than 5, do not change the previous place number.

❷ To round a number to significant places, count the number of digits from the first non-zero digit, starting from the left then round as above.

❸ To estimate the approximate answer to a calculation, round each number to **one** significant figure (1 s.f.).

Key terms

Decimal places

Significant figures

Approximation

Estimate

Look out for

When identifying significant figures, remember the first two 0s here are not significant; the number 1 is the first significant figure.

Worked examples

a Write 4.754

   i correct to 1 decimal place

   ii correct to 3 significant figures.

   Answers

   ❶ **i** 4.754 = 4.8

   ❷ **ii** 4.754 = 4.75

b Write 0.01278

   i correct to 2 decimal places,

   ii correct to 2 significant figures.

   Answers

   ❶ **i** 0.01278 = 0.01

   ❷ **ii** 0.01278 = 0.013

c Write down an estimate for the value of

   i 1026

   ii 0.498

   Answers

   ❸ **i** 1026 = 1000

   ii 0.498 = 0.5

❶ⓐ the next number after the required number of decimal places

❷ⓐ the next number after the required number of significant figures

❶ⓑ the next number after the required number of decimal places

❷ⓑ the next number after the required number of significant figures

Exam tip

The size of the number does not change.

Look out for

A common mistake is to write 0.498 = 0

Exam-style questions

1 The dimensions of a rectangle are $4.87\,\text{cm} \times 2.35\,\text{cm}$.

   Work out the area of this rectangle.

   Give your answer correct to 2 decimal places. **[2]**

2 The length of a piece of string is given as 12 cm correct to 2 significant figures.

   Write down the least possible actual length of this piece of string. **[1]**

3 Find an estimate for the value of $\frac{4.83 \times 204}{0.51}$ **[2]**

Exam tip

Round each number correct to **one** significant figure.

CHECKED ANSWERS

Limits of accuracy

Rules

1. Given a degree of accuracy for a number, to find a lower bound, write down the midpoint of the given number and the number with one degree of accuracy less.
2. Given a degree of accuracy for a number, to find an upper bound, write down the midpoint of the given number and the number with one degree of accuracy more.

Worked examples

a The length of a football pitch is measured as 120 yards to the nearest yard.

Write down the
i   lower bound
ii upper bound of this length.

Answers

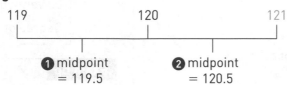

119   120   121

❶ midpoint = 119.5 ❷ midpoint = 120.5

i   lower bound = 119.5 yards
ii upper bound = 120.5 yards

b The volume of a bottle is 85 cm³ correct to the nearest 5 cm³. Write down the **i** lower bound and **ii** upper bound of this volume.

Answers

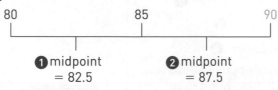

80   85   90

❶ midpoint = 82.5 ❷ midpoint = 87.5

i   lower bound = 82.5 cm³
ii upper bound = 87.5 cm³

Key terms

Upper bound

Lower bound

Degree of accuracy

Exam tip

Draw a diagram to show the numbers below and above that given.

Look out for

5 cm³ is the degree of accuracy, so the scale must go 5 cm³ below and 5 cm³ above.

Remember

The upper bounds are boundaries, not values that the quantity could actually equal; so do **not** write, 120.49... or 87.49... in i and ii.

Exam-style questions

1. The dimensions of the top of a table are given as 2.3 m × 1.2 m measured correct to 1 decimal place. Write down the
   a lower bound
   b upper bound of these dimensions. **[2]**

2. Mo runs a distance of 2.5 km measured correct to the nearest 10 metres. Find the lower bound of Mo's run. **[1]**

3. The distance of Milly's house to her grandfather's house is 190 miles measured to the nearest 10 miles. It took Milly **exactly** three hours to drive to her grandfather's house. Milly says 'my average speed was 60 mph'. Could Milly be right? Explain your answer. **[3]**

Exam tip

Remember: speed = $\dfrac{\text{distance}}{\text{time}}$

CHECKED ANSWERS

Calculating with lower and upper bounds

HIGH

Rules

❶ Use both lower bounds (or upper bounds) when finding the lower bound (or upper bound) of the sum or product of two quantities.

❷ When finding the lower bound of a quotient or the difference between two quantities, work out the lower bound divided by the upper bound or the lower bound minus the upper bound.

❸ When finding the upper bound of a quotient or the difference between two quantities, work out the upper bound divided by the lower bound or the upper bound minus the lower bound.

❹ The degree of accuracy of a calculation is given by rounding the results to common significant figures.

Worked examples

a The dimensions of the top of a rectangle are given as $12.0\,\text{cm} \times 8.0\,\text{cm}$ measured correct to 1 decimal place. Find the bounds of:

  i the perimeter

  ii the area of this rectangle.

  iii Find the area to an appropriate degree of accuracy.

Answers

  i lower bound of length $= 11.95\,\text{cm}$

  upper bound of length $= 12.05\,\text{cm}$

  lower bound of width $= 7.95\,\text{cm}$ upper bound of width $= 8.05\,\text{cm}$

  lower bound of perimeter $= 2 \times (11.95 + 7.95) = 39.8\,\text{cm}$ ❶

  upper bound of perimeter $= 2 \times (12.05 + 8.05) = 40.2\,\text{cm}$ ❶

  ii lower bound of area $= 11.95 \times 7.95 = 95.0025\,\text{cm}^2$ ❶

  upper bound of area $= 12.05 \times 8.05 = 97.0025\,\text{cm}^2$ ❶

  iii The degree of accuracy $= 100\,\text{cm}^2$ ❹ ← (since both 95.0025 and 97.0025 round to 100)

b $D = 28$ miles correct to the nearest 2 miles. $T = 15$ minutes correct to the nearest minute. Work out the upper bound of S, where $S = \dfrac{D}{T}$.

Answer

upper bound of $S = \dfrac{\text{upper bound of } D}{\text{lower bound of } T}$ ❸ $= 29 \div 14.5 = 2$ miles/minute

Key terms

Upper bound

Lower bound

Degree of accuracy

Exam tip

Draw a diagram to find lower and upper bounds, as limits of accuracy.

Look out for

A change in units; mph could have been required here.

Exam-style questions

1 The area of a circle is estimated to be $54\,\text{cm}^2$ to the nearest $2\,\text{cm}^2$. The value of π is taken to be 3.14 to 2 decimal places.

Work out the length of the radius to an appropriate degree of accuracy. **[4]**

2 Gerry can hold his breath for 58.5 seconds measured to the nearest half a second. Mary can hold her breath for 1 minute 2.5 seconds measured to the nearest half a second.

Work out the least possible difference between these two times. **[3]**

Look out for

A common mistake is to work with all upper (lower) bounds when finding an upper (lower) bound of a calculation. Use rules ❶, ❷ and ❸.

CHECKED ANSWERS

Reverse percentages

Rules

1. If the final value is the result of a percentage increase, add 100% to the percentage increase and divide the final value by this new percentage.
2. If the final value is the result of a percentage decrease, subtract the percentage decrease from 100% and divide the final value by this new percentage.

Worked examples

a In a sale the price of a TV is reduced by 15%.
If the sale price is £544, work out the original price of the TV.

Answer

The price is a reduction so: 100% − 15% = 85% (which is $\frac{85}{100}$).

❷ $544 \div \frac{85}{100} = 544 \times \frac{100}{85} = £640$

b N is increased by 80%. Its value is now 126.

What was the value of N?

Answer

This is an increase so: 100% + 80% = 180% (which is $\frac{180}{100} = 1.8$)

N = 126 ÷ 1.8 = 70

Key terms

Percentage increase

Percentage decrease

Remember

To divide by a fraction, invert the fraction then multiply by it.

This is the multiplier of the increase

Exam tip

If using a multiplier, show how you get it.

Exam-style questions

1 Ismail bought a smart phone and a laptop. The total cost, including VAT at a rate of 20%, was £684. The price of the smart phone excluding VAT was £250.

What was the price of the laptop excluding VAT? **[3]**

2 David keeps bees. In 2015 he had 6400 bees. This was an increase of 27.5% on 2014.

David said he had fewer than 5000 bees in 2014. Is David right? **[3]**

3 Ben has changed his job. His new salary is 5% less than before. Ben's wife Jane has just had an 8% increase in her salary. Ben's salary is now £26 500 per year. Jane's salary is now £22 000 per year.

Are their total earnings better or worse now? **[4]**

Look out for

Be clear if the answer is going to be greater than or less than the original value.

Look out for

A common mistake is to find the percentage of the final amount and subtract or add depending upon whether the original value will be less or greater.

CHECKED ANSWERS

Repeated percentage increase/decrease

LOW

Rules

❶ Find the increase (decrease) after one period of time and add (subtract) this to the original amount. The percentage increase (decrease) is then applied to this total amount and a new total found. This continues for the required number of repetitions.

❷ If an increase or decrease in percentage is repeated n times, the compounded value can be found using the multiplier raised to the power of n.

Key term

Compound interest

Worked examples

a Tim invests £4000 in a savings account. Compound interest is paid at a rate of 3.5% per annum. How much will Tim have in his account after 4 years?

Answer

❶ 3.5% of $4000 = \frac{3.5}{100} \times 4000 = 140$

Total after 1 year $= 4000 + 140 = 4140$

❶ 3.5% of $4140 = \frac{3.5}{100} \times 4140 = 144.90$

Total after 2 years $= 4140 + 144.90 = 4284.90$

❶ 3.5% of $4284.90 = \frac{3.5}{100} \times 4284.90 = 149.97$

Total after 3 years $= 4284.90 + 149.97 = 4434.87$

❶ 3.5% of $4434.87 = \frac{3.5}{100} \times 4434.87 = 155.22$

Total after 4 years $= 4434.87 + 155.22 = 4590.09$

❷ Multiplier $= \frac{100 + 3.5}{100} = 1.035$

$n = 4$ since it is a 4-year period

Total after 4 years $= 4000 \times 1.035^4 = 4590.09$

Exam tip

1.035^4 is found using the y^x button on the calculator. Some calculators do not have a y^x.

Exam tip

Method ❷ is clearly a more direct method.

b The population of birds in a bird sanctuary is 4500. It is estimated that the population will decrease at a rate of 12% each year for the next 3 years. What is the expected population after the next 3 years?

Answer

Multiplier $= \frac{100 - 12}{100} = 0.88$

❷ Expected population $= 4500 \times 0.88^3 = 3066$

Look out for

Be clear if the answer is going to be greater than or less than the original value.

Exam-style questions

1 A bank pays interest at a rate of 4.5% for the first year and 2% for each subsequent year. Tess invests £35 000 for 5 years. Work out the total interest paid at the end of 5 years. **[3]**

2 Chris bought a new car for £17 500. It is estimated that the car will depreciate in value by 20% in the first year, 15% in the second year and 12% pa for the next 3 years. Chris says that after 5 years the value of the car will be greater than half the cost of the car. Is Chris right? **[4]**

3 Jose has just opened a new restaurant. He predicts that his profits will increase by 12.5% every 6 months. How many years will have passed before Jose's profits are double what they were after the first 6-month period? **[3]**

Exam tip

Use multipliers wherever possible in this exercise.

CHECKED ANSWERS

Growth and decay

Rules

1. If an increase or decrease in percentage is repeated n times, the compounded value can be found using the multiplier raised to the power n.
2. To find the repeated percentage increase (or decrease) given n, the number of repetitions, divide the final amount by the original amount and then find the nth root. The nth root is the multiplier. The percentage change is then found by multiplying by 100 and then either adding or subtracting from 100.
3. To find the number of repetitions in a growth or decay problem, divide the final amount by the original amount to give a scale factor. A power of the percentage-derived multiplier is then equated to this and solved to give the number of repetitions.

Key terms

Exponential growth and decay

Worked examples

a The population of a rare species of spider decreases at a rate of 60% per year. 12 years ago the population of this spider was 6 million. What is it now?

Answer

Decrease of 60% p.a., so multiplier $= 0.4\left(\frac{100-60}{100}\right)$

Population after 12 years $= 6\,000\,000 \times 0.4^{12} = 100$ ①

Look out for

The answer here must be a whole number.

b The value of a painting increased over a 6 year period from £25 000 to just over £33 500. Work out the percentage increase each year.

Answer

Multiplier $= \sqrt[6]{33\,500 \div 25\,000} = 1.049\,987... = 1.05$

Percentage change $= 1.05 \times 100 - 100 = 105 - 100 = \textbf{5\%}$ ②

This can be approximated to 1.05 since the value was **over** £33 500.

c A government bond pays compound interest at a rate of 10% per year. How many years would it take for an investment of £1000 to double in value?

Answer

Rate of interest is 10% so multiplier $= 1.1$

$1000 \times 1.1^x = 2000$

So $1.1^x = 2$, $1.1^7 = 1.948$ and $1.1^8 = 2.14$. So $x = 8$ years. ③

Exam tip

Trial and improvement can be applied here.

Exam-style questions

1 The temperature of a cup of tea is 85 °C. The temperature drops at a rate of 25% per minute.

Work out the temperature of the tea after 6 minutes. Give your answer correct to 3 significant figures. **[2]**

2 A political party of a country has 60 seats in its parliament. It is expected that the number of seats will increase by 12% each year.

How many years will it take before the number of seats has increased by 50%? **[3]**

CHECKED ANSWERS

Mixed exam-style questions

1. The length of a rectangle is 8.364 cm. The width of the rectangle is 5.549 cm.
   Tony says that the area of the rectangle is least when the length and width are rounded to 1 significant figure.
   Noreen says it is least when the length and width are rounded to 2 significant figures.
   Waqar says it is least when the length and width are rounded to 3 significant figures.
   Who is right? [4]

2. Saturn is 1.25×10^9 km from Earth. Venus is 4.14×10^7 km from Earth.
   a How many times is Saturn further from Earth than Venus? [2]
   b How many miles is Saturn from Earth? [2]

3. Prove algebraically that the recurring decimal $5.7\dot{2}$ can be written as the fraction $5\frac{13}{18}$. [3]

4. The dimensions of a rectangular piece of paper are given as 30 cm × 18 cm measured correct to the nearest centimetre.
   Could the area of this piece of paper be greater than 550 cm²?
   Explain your answer. [3]

5. Heddwen drives to her friend's house.
   The distance Heddwen drives is 3.6 miles measured correct to the nearest tenth of a mile.
   It takes Heddwen 7.2 minutes measured correct to one decimal place.
   The speed limit throughout Heddwen's journey was 30 mph.
   Could Heddwen have always been driving under the speed limit?
   Explain your answer. [4]

6. From Monday, the price of a TV was reduced by 20% in a sale.
   On Wednesday the TV was reduced by a further 10%.
   Alan bought the TV on Wednesday for £604.80.
   What was the price of the TV before the sale? [4]

7. At the end of 2014, the population of Summerstown was 15 500.
   At the end of 2015 the population had increased by 500.
   The percentage increase in the population of Summerstown was predicted to increase at a constant rate for the next 9 years.
   Dafydd said that this means the population will have increased by 5000 after these 10 years.
   Explain fully why Dafydd is wrong and use this information to correctly predict the population of Summerstown at the end of 2024. [4]

8. Lois breeds tropical fish.
   After 4 months, she has 1200 fish.
   After 6 months she has 1500 fish.
   Assuming that the number of fish have increased exponentially, how many fish did Lois have after 1 year? [5]

Working with proportional quantities

Rules

❶ To use the unitary method, find out what proportion is just **one** part of the whole amount,

❷ Then multiples of that can be found.

Worked examples

a 12 identical books cost £23.88.

Work out the cost of 5 of these books.

Answer

❶ 23.88 ÷ 12 = £1.99 per book ←————————————

❷ 5 books cost £1.99 × 5 = £9.95

The value of **one** unit.

b Work out which is the better value for these bags of potatoes; 6 kg for £8.16 or 11 kg for £15.18.

Answer

❶ £8.16 ÷ 6 = £1.36 per kg

❶ £15.18 ÷ 11 = £1.38 per kg

So 6 kg for £8.16 is the better value.

Key terms

Ratio

Proportion

Multiples

Exam tip

Always give answers in a sentence supported by working.

Exam-style questions

1 8 pens cost £5.20.

Work out the cost of 15 of these pens. **[2]**

2 Jay buys three portions of chips and two pies for £6.45. Mandy buys five pies for £6.

How much does one portion of chips cost? **[3]**

3 Here are the ingredients to make 40 biscuits.
600 g of butter, 300 g of sugar and 900 g of flour.
Mrs Bee has the following ingredients in her cupboard.
1.5 kg of butter, 1 kg of sugar and 2 kg of flour.

Work out the greatest number of these biscuits that Mrs Bee can make. **[4]**

Exam tip

Always work out the value of **one** part.

Exam tip

Explain why this is the greatest number.

CHECKED ANSWERS

The constant of proportionality

LOW

Rules

❶ To work out a constant of proportionality of two variables which are in direct proportion, divide one variable by the other.

❷ To derive a formula describing the relationship between two variables, find the constant of proportionality and then substitute into the relationship.

Worked examples

a The table of values shows the miles (D) travelled by a car using G gallons of petrol.

D miles	120	240	480	720	1200
G gallons	5	10	20	30	50

Write down a formula connecting D and G.

Answer

D ∝ G, so D = kG where k is the constant of proportionality.

❶ 120 ÷ 5 = 24; 240 ÷ 10 = 24; 480 ÷ 20 = 24; 720 ÷ 30 = 24; 1200 ÷ 50 = 24

$k = 24$ is the constant of proportionality.

❷ D = 24G

b y is directly proportional to x.
When $x = 5$, $y = 11$.
Work out the value of x when $y = 30$.
Give your answer correct to 1 decimal place.

Answer

$y ∝ x$, so $y = kx$

When $x = 5$, $y = 11$, so $11 = k \times 5$

$k = 11 ÷ 5 = 2.2$ ❶

$y = 2.2x$ ❷

When $y = 30$, $30 = 2.2x$

$x = 30 ÷ 2.2 = 13.6$

Exam tip

Remember: '∝' means 'is proportional to'

Key terms

Direct proportion

Constant of proportionality

Exam tip

Step 1: use given information to find k.

Step 2: use value of k to write down formula.

Step 3: use formula to find the required unknown.

Exam-style questions

1 The table of values shows amounts of money in pounds (£P) and their equivalent values in euros (€E)

Pounds (£P)	2.00		6.00		10.00
Euros (€E)	2.70	5.40		10.80	13.50

a   Write down the missing figures from the table. **[1]**
b   Write down a formula for E in terms of P. **[2]**
c   What does the constant of proportionality represent in this formula? **[1]**

2 H is directly proportional to t.
When $t = 5.6$, $H = 14$
Work out the value of H when $t = 35$ **[3]**

CHECKED ANSWERS

Working with inversely proportional quantities

REVISED ☐

LOW

Rules

1 To work out a constant of proportionality of two variables which are inversely proportional to each other, multiply one variable by the other.

2 To derive a formula describing the relationship between two variables, find the constant of proportionality and then substitute into the relationship.

Worked examples

a A sum of money is divided equally between N people so that each person gets $£p$.
  i If 30 people each get £25, write down a formula for N in terms of p.
  ii What does the constant of proportionality represent?

Answers

i N is inversely proportional to p, so $N = \frac{k}{p}$

  1 $k = 30 \times 25 = 750$, so $N = \frac{750}{p}$ **2**

ii The constant of proportionality, 750, is the amount of money shared.

b W is inversely proportional to d. When $d = 5$, $W = 120$. Work out the value of d when $W = 600$.

Answer

$W \propto \frac{1}{d}$, so $W = \frac{k}{d}$, when $d = 5$, $W = 120$, so $120 = k \div 5$

$k = 120 \times 5 = 600$ **1**

$W = \frac{600}{d}$

When $W = 600$, $600 = 600 \div d$

$d = 600 \div 600 = 1$

Exam tip

Remember:
'$\propto$' means 'is proportional to'
'inverse' means '1 over' or 'reciprocal'

Key terms

Inverse proportion

Constant of proportionality

Exam tip

Step 1: use given information to find k.

Step 2: use value of k to write down formula.

Step 3: use formula to find the required unknown.

Note: this is the **same** process as for direct proportion.

Exam-style questions

1 It takes 8 men 25 days to build a wall. It takes 10 men 20 days to build an identical wall.
  a Is this an example of direct or inverse proportion? You must explain your answer. **[1]**
  b How many days would it take 4 men to build the wall? **[2]**
     (Note: All of the men work at the same rate.)

2 y is inversely proportional to x. The table of values has just one error. What is it? **[2]**

x	1.2	1.5	25	30	120
y	250	200	120	10	2.5

3 P is inversely proportional to s.
  When $s = 20$, $P = 100$. Work out the value of s when $P = 200$ **[2]**

CHECKED ANSWERS ☐

Formulating equations to solve proportion problems

REVISED

HIGH

Rules

1. To derive an equation for direct proportion between y and a function of x, $f(x)$ use the formula $y = kf(x)$.
2. To derive an equation for inverse proportion between y and a function of x, $f(x)$ inverse proportion, use the formula $y = \frac{k}{f(x)}$.
3. In each case, k, the constant of proportionality, must be found.

Key terms

Direct / inverse proportion

Constant of proportionality

Worked examples

a The volume of a sphere is directly proportional to the cube of its radius. A ball has a radius, r, of 1.5 cm and a volume, V, of $4.5\pi\,\text{cm}^3$. Write an equation giving V in terms of r.

Answer

$V = k \times r^3$ ❶

When $r = 1.5$, $V = 4.5\pi$, so $4.5\pi = k \times 1.5^3$

$k = 4.5\pi \div 1.5^3 = 1.333\dot{3}\pi = \frac{4}{3}\pi$ ❸

So $V = \frac{4}{3}\pi r^3$

b T is inversely proportional to the square root of w.

When $w = 1.96$, $T = 25$.

Write an equation giving T in terms of w.

Answer

$T = k \times \frac{1}{\sqrt{w}}$ ❷

When $w = 1.96$, $T = 25$. So, $25 = k \times \frac{1}{\sqrt{1.96}} = \frac{k}{1.4}$

$k = 25 \times 1.4 = 35$ ❸

So $T = \frac{35}{\sqrt{w}}$

Exam tip

Remember: 'is ... proportional to' is replaced by '$= k \times ...$'

Exam tip

Read the question carefully; many errors are made by writing an incorrect first step, for example, $V = k \times r^3$ and $T = k \times \frac{1}{\sqrt{w}}$.

Exam-style questions

1. A pebble is dropped down a well. The distance travelled, D metres, by the pebble is directly proportional to the square of the time of travel, t seconds.
   a Derive a formula for D in terms of t if a pebble falls a distance of 20 metres in 2 seconds. **[3]**
   b Work out how many seconds it would take the pebble to fall a distance of 80 metres. **[2]**

2. y is inversely proportional to the cube root of x.
   Find the missing values in this table. **[4]**

x	1	8	64	216	
y				400	240

3. The light intensity, I units, from a light source is inversely proportional to the square of the distance d cm from the light source. At a distance of 10 cm from a light source the intensity is 64 units.

   Work out the intensity at a distance of 2 cm from the light source. **[4]**

CHECKED ANSWERS

Index notation and rules of indices

Rules

❶ $a \times a \times a \times \dots \times a$ (m times) is written a^m.

❷ To multiply numbers written in index form, add the powers together. $a^m \times a^n = a^{m+n}$

❸ To divide numbers written in index form, subtract the powers. $a^m \div a^n = a^{m-n}$

❹ To raise a number written in index form to a given power, multiply the powers together. $(a^m)^n = a^{mn}$

Key terms

Index

Indices

Powers

Exam tip

A common mistake is to multiply powers instead of adding in a product.

Exam tip

A common mistake is to divide powers instead of subtracting in a quotient.

Look out for

Follow the rules of BIDMAS and work out the calculation inside the brackets first.

Worked examples

a Write $7 \times 7 \times 7 \times 7 \times 7$ in index form.

Answer

❶ $7 \times 7 \times 7 \times 7 \times 7 = 7^5$

b Write $\left(\frac{2^3 \times 2^4}{2^5}\right)^3$ as a power of 2.

Answer

$\left(\frac{2^3 \times 2^4}{2^5}\right)^3 = \left(\frac{2^7}{2^5}\right)^3$ since $2^3 \times 2^4 = 2^{3+4} = 2^7$ ❷

$= (2^2)^3$ since $2^7 \div 2^5 = 2^{7-5} = 2^2$ ❸

$= 2^6$ since $(2^2)^3 = 2^{2 \times 3} = 2^6$ ❹

Exam-style questions

1 a Write $10 \times 10 \times 10 \times 10$ in index notation. **[1]**

  b Use your calculator to work out the value of 8^5. **[1]**

2 $x = 8 \times 2^4$ $y = 4^2 \times 16$

Work out the value of xy. Give your answer as a power of 2 **[3]**

3 Tom is trying to work out the value of $\frac{10^4 \times 10^5}{10 \times 10^2}$

Tom writes $\frac{10^4 \times 10^5}{10 \times 10^2} = \frac{10^{20}}{10^2} = 10^{10} = 100$.

Write down each of the mistakes that Tom has made. **[4]**

Fractional indices

Rules

① The denominator of a fractional index is the root; for example, in $x^{\frac{m}{n}}$, n is the nth root.

② The numerator of a fractional index is the power; for example, in $x^{\frac{m}{n}}$, m is the mth power.

③ $x^{\frac{m}{n}}$ can be written as $\left(\sqrt[n]{x}\right)^m$, or

④ $\sqrt[n]{x^m}$.

⑤ Negative indices means the reciprocal; for example, $x^{-\frac{m}{n}} = \dfrac{1}{\left(\sqrt[n]{x}\right)^m}$.

Key terms

Fractional index

Power

Root

Reciprocal

Worked examples

a Write down the value of $243^{\frac{1}{5}}$.

Answer

In $243^{\frac{1}{5}}$, the '5' means the 5th root of 243 ①

$243^{\frac{1}{5}} = 3$

b Write down the value of $27^{-\frac{2}{3}}$.

Answer

$27^{-\frac{2}{3}} = \dfrac{1}{27^{\frac{2}{3}}} = \dfrac{1}{\left(\sqrt[3]{27}\right)^2} = \dfrac{1}{3^2} = \dfrac{1}{9}$

      ⑤   ③

Or,

$27^{-\frac{2}{3}} = \dfrac{1}{27^{\frac{2}{3}}} = \dfrac{1}{\left(\sqrt[3]{27}\right)^2} = \dfrac{1}{\sqrt[3]{729}} = \dfrac{1}{9}$

      ⑤   ④

c $x^n = \sqrt{x} \div \dfrac{1}{x^3}$. Find the value of n.

Answer

$x^n = \sqrt{x} \div \dfrac{1}{x^3} = x^{\frac{1}{2}} \div x^{-3} = x^{\frac{1}{2}-(-3)} = x^{3.5}$, so $n = 3.5$.

      ①   ⑤

Exam tip

It is generally advised to work out the root before the power since the calculations (for example, $\sqrt[3]{27} = 3$ and $3^2 = 9$ rather than $27^2 = 729$ and $\sqrt[3]{729} = 9$) are easier.

Exam-style questions

1 Find the value of:

  a $16^{\frac{3}{4}}$ **[1]**

  b $8^{-\frac{2}{3}}$ **[2]**

2 Write these numbers in order of size. Start with the smallest value. **[3]**

$16^{-\frac{1}{4}}$ 27^0 $81^{\frac{3}{4}}$ $25^{-\frac{1}{2}}$ $\dfrac{1}{4^{-\frac{1}{2}}}$

3 $5^n = \dfrac{5 \times \sqrt[3]{5^4}}{\left(\sqrt[4]{5}\right)^2}$

Find the value of n. **[3]**

Surds

Rules

1 To simplify a surd, look for factors that are square numbers and then factorise by taking out the square root.

2 To rationalise the denominator (in the form $\sqrt{n}$) of a fraction, multiply the fraction by $\frac{\sqrt{n}}{\sqrt{n}}$.

3 To rationalise the denominator (in the form $\sqrt{n}+m$) of a fraction, multiply the fraction by $\frac{\sqrt{n}-m}{\sqrt{n}-m}$.

Key terms

Rational number

Irrational number

Rationalise denominator

Worked examples

a Simplify $\sqrt{180}$.

Answer

$\sqrt{180} = \sqrt{4 \times 5 \times 9} = \sqrt{4} \times \sqrt{5} \times \sqrt{9} = 2 \times \sqrt{5} \times 3 = 6\sqrt{5}$ **1**

b Rationalise the denominator of $\frac{4}{\sqrt{3}}$.

Answer

$\frac{4}{\sqrt{3}} \times \frac{\sqrt{3}}{\sqrt{3}} = \frac{4\sqrt{3}}{\sqrt{3} \times \sqrt{3}} = \frac{4\sqrt{3}}{3}$ **2**

c Rationalise the denominator of $\frac{2}{1-\sqrt{5}}$.

Answer

$\frac{2}{1-\sqrt{5}} \times \frac{1+\sqrt{5}}{1+\sqrt{5}} = \frac{2(1+\sqrt{5})}{(1-\sqrt{5})(1+\sqrt{5})} = \frac{2+2\sqrt{5}}{1-(\sqrt{5})^2} = \frac{2+2\sqrt{5}}{4}$ or $\frac{1+\sqrt{5}}{2}$ **3**

Exam tip

Remember: $\frac{\sqrt{3}}{\sqrt{3}} = 1$, so multiplying by this does not change the value.

Exam tip

This uses the theory of the difference of two squares: $(a-b)(a+b) = a^2 - b^2$.

Exam-style questions

1 Rationalise the denominator of:

a $\frac{\sqrt{3}}{\sqrt{7}}$ **[1]**

b $\frac{2}{\sqrt{5}-2}$ **[1]**

2 The first two terms of a geometric progression are 1 and $\frac{2}{\sqrt{3}}$.
Work out the 6th term of this sequence, rationalising the denominator of your answer. **[3]**

3 Work out the volume of a cube of side $(1 + \sqrt{2})$ cm.
Give your answer in the form $a + b\sqrt{2}$ **[3]**

CHECKED ANSWERS

Mixed exam-style questions

1 y is directly proportional to x and x is directly proportional to z.
   a Prove that y is directly proportional to z. [2]
   b When $z = 8$, $x = 40$ and $y = 160$.
   Work out the value of y when $z = 2.5$. [2]

2 W is inversely proportional to t.
   W is directly proportional to s.
   a Write down the relationship between t and s. [1]
   b $W = 8s$ and $t = 10$ when $W = 4$.
   Find the value of t when $s = 1.5$. [3]

3 Maria wants to find the height, h metres, of a cliff. She drops a stone from the top and measures the time, t seconds, it takes to reach the ground below. The height is directly proportional to the square of the time.
   If it takes 2 seconds for the stone to fall 20 metres, work out the height of the cliff if Maria's stone takes 7.5 seconds to reach the ground below. [3]

4 $2^3 \times 2^{2x-1} = 8^{-1}$
   Find the value of x. [3]

5 Find the difference between the LCM and the HCF of 90 and 84. [4]

6 Find the value of:
   a $100^{-\frac{1}{2}}$ [1]
   b $16^{\frac{1}{4}} \times 125^{-\frac{1}{3}}$ [2]

7 Rationalise the denominator of $\frac{10}{\sqrt{5}}$. [2]

Algebra: pre-revision check

Check how well you know each topic by answering these questions. If you get a question wrong, go to the page number in brackets to revise that topic.

1 a Simplify these.

  i $a^4 \times a^6$

  ii $\dfrac{x^8}{x^5}$

  iii $\dfrac{12e^6f^7}{8e^9f^5}$

  b Expand and simplify these.

  i $(t + 2)(t + 5)$

  ii $(v - 7)(v + 5)$

  iii $(y - 6)(y - 5)$ (Page 21)

2 This formula is used to find the distance, s, travelled by an object.

$$s = ut + \tfrac{1}{2}at^2$$

  a Find the value of s when $u = 5$, $t = 4$ and $a = 10$.

  b Make a the subject of the formula.
(Page 22)

3 Prove that the sum of the three consecutive numbers $(n - 1)$, n and $(n + 1)$ is a multiple of 3.   (Page 23)

4 Simplify these.

  a $\sqrt{\dfrac{a^6b^8}{c^4}}$

  b $\dfrac{a^3b^{\frac{5}{2}}c^{\frac{3}{4}}}{a^{\frac{3}{2}}b^4c^{\frac{1}{2}}}$ (Page 24)

5 Solve $\dfrac{5x}{x + 5} - \dfrac{3}{x - 2} = 5$.   (Page 25)

6 Here is a formula used to find the final speed, v, of an object:

$$v^2 - u^2 = 2as$$

  a Find the value of v when $u = 20$, $a = 5$ and $s = 80$.

  b Make x the subject of the formula $y = \dfrac{x + 5}{3 - 2x}$.
(Page 26)

7 The nth term of quadratic sequence is $n^2 + 5$.
The mth term of a different quadratic sequence is $80 - 2m^2$.
Find the number that is in both sequences.
(Page 28)

8 Here is a geometric sequence.

    5   15   45   135   ...

  a Find the common ratio.

  b Find the 10th term of the sequence.
(Page 29)

9 Find the nth term of this quadratic sequence:

    3   6   13   24   39   ... (Page 29)

10 Find the equation of a straight line graph that passes through the point $(-3, 3)$ and is parallel to the line $x + 2y = 8$.   (Page 32)

11 a Sketch the graph of the quadratic function $y = x^2 - 4x + 3$ for values of x from 0 to 5.

  b Write down the roots of the equation $x^2 - 4x + 3 = 0$.

  c Write down the line of symmetry of the graph.   (Page 33)

12 On a co-ordinate grid drawn with values of x from -3 to $+3$ and values of y from -10 to $+30$:

  a draw the graph of $y = x^3 + x^2 - 3x$

  b find the values of x when $x^3 + x^2 - 3x = 0$.
(Page 33)

13 Find the equation of the line that is perpendicular to $y = 2x + 3$ and passes through the point $(4, 3)$.
(Page 34)

14 a Write down the inequality shown on this number line.

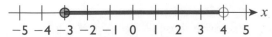

  b Solve these inequalities.

  i $2x + 5 < 9$

  ii $24 + 2t > 30 - 3t$

  iii $5(y - 3) \leqslant 3y - 6$ (Page 41)

15 Solve this pair of simultaneous equations.

$5x + 2y = 8$

$2x - y = 5$ (Page 42)

16 Here is the graph of the line $y + 2x = 3$.

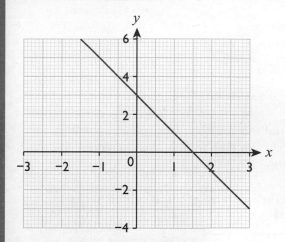

Find graphically the solution to the simultaneous equations:

$y + 2x = 3$

$y - 2x = 1$ (Page 43)

17 On a co-ordinate grid with values of x from -3 to $+3$ and y from -4 to $+6$, show the region defined by these inequalities:

$x + y > -1$ $y \leqslant 1 - 2x$ $y \leqslant x + 3$ (Page 44)

18 a Expand and simplify these.

   i $(x + 4)(x - 5)$

   ii $(y + 8)(y - 8)$

   iii $(6 - a)(a + 6)$

b Factorise these.

   i $x^2 + 7x + 12$

   ii $e^2 - 3e - 10$

   iii $b^2 - 25$ (Page 45)

19 Solve these equations.

a $x^2 - 5x + 6 = 0$

b $x^2 - 2x = 15$

c $p^2 - 49 = 0$ (Page 46)

20 a Factorise these.

   i $4x^2 + 4x - 3$

   ii $9b^2 - 64$

b Solve these.

   i $3x^2 + 11x - 20 = 0$

   ii $\dfrac{2x}{x+3} - \dfrac{x}{x+2} = 1$

c Simplify $\dfrac{3x+2}{3x^2 - 13x - 10}$.   (Page 46)

21 Solve these. Give your answer to 3 decimal places.

a $2x^2 + 2x - 1 = 0$

b $\dfrac{4x}{x-3} - \dfrac{x}{x+1} = 2$ (Page 47)

22 The graph shows the speed of a car as it accelerates away from rest.

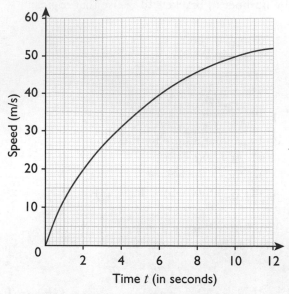

a   Find the acceleration when $t = 4$.

b   Find the average acceleration between $t = 2$ and $t = 10$.   (Page 50)

23 Here is the graph of $y = f(x)$.

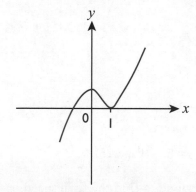

Sketch the graph of:

a $f(-x)$

b $f(x - 2)$.   (Page 52)

24 For the graph in Question 22, find, using 5 strips of width 2 seconds, the distance travelled in the first 10 seconds.   (Page 53)

Simplifying harder expressions and expanding two brackets

LOW

Rules

1. Index law for multiplying numbers or variables raised to a power is
$$a^n \times a^m = a^{n+m}$$
2. Index law for dividing numbers of variables raised to a power is
$$a^n \div a^m = a^{n-m}$$
3. Index law for raising a variable written as a power to a power is
$$(a^n)^m = a^{n \times m}$$
4. When you expand a pair of brackets you multiply every term in the second bracket by every term in the first bracket.

Worked examples

a Work out

  i $3x^4 \times 5x^6$ **ii** $\dfrac{12y^7}{4y^3}$ **iii** $(2f^3)^5$ **iv** $\dfrac{6a^6b^4 \times 4a^2b^5}{12a^5b^6}$

Answers

Deal with the whole numbers in front of the variables first.

$3 \times 5 = 15$ $12 \div 4 = 3$ $2^5 = 32$ $6 \times 4 \div 12 = 2$

Now deal with the powers or indices.

1 $15x^{4+6}$ **2** $3y^{7-3}$ **3** $32f^{3 \times 5}$ **1 2** $2a^{6+2-5}b^{4+5-6}$

  $= 15x^{10}$ $= 3y^4$ $= 32f^{15}$ $= 2a^3b^3$

b Expand and simplify
$(x - 4)(x + 6)$

Answer

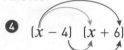

4 $(x - 4) \ (x + 6)$

$x \times x = x^2;\ x \times +6 = 6x$
$-4 \times x = -4x;\ -4 \times +6 = -24$ **OR**
$x^2 + 6x - 4x - 24$
$x^2 + 2x - 24$

$\times$	x	$+6$
x	x^2	$+6x$
-4	$-4x$	-24

$x^2 + 6x - 4x - 24$
$x^2 + 2x - 24$

Look out for

Any number or variable with a power or index of 0 is always 1, e.g. $a^0 = 1$; $25^0 = 1$.

Key terms

Power

Index

Variable

Bracket

Expand

Simplify

Exam-style questions

1 Work out

  a $\dfrac{12(y^3)^7}{16y^{15}}$ **[2]**

  b $\dfrac{3a^5b^3 \times 5a^4b^5}{12(ab)^6}$ **[2]**

2 This shape is made from a large rectangle and a blue square.

  Explain why the area of the red shape is $a^2 + 11a + 15$ **[3]**

Exam tip

Remember that powers and indices are the same.

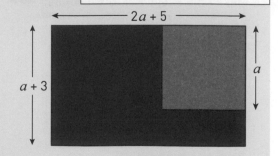

CHECKED ANSWERS

Using complex formulae and changing the subject of a formula

Rules

1. You can replace the variables or letters in a formula with positive and negative numbers.
2. Use BIDMAS and rules for dealing with adding, subtracting, multiplying and dividing positive and negative numbers to find the value of the missing letter.
3. Use inverses to change the subject of a formula so that the required variable or letter is on its own on one side of the formula or equation.

Worked examples

a Work out the value of p when $a = 2.5$, $b = -2$ and $c = -5$

$$p = \frac{2(a^2 - b^2)}{5 - 3c}$$

Answer

Firstly replace the variables or letters with their values.

$$p = \frac{2((2.5)^2 - (-2)^2)}{5 - 3 \times (-5)}$$ **①**

Then use BIDMAS and the positive and negative sign rules to get:

$$p = \frac{2 \times 6.25 - 2 \times (+4)}{5 + 15}$$ $(2.5^2 = 6.25, (-2)^2 = +4$ and $-3 \times -5 = +15)$ **②**

Then work it all out

$$p = \frac{12.5 - 8}{20} = 4.5 \div 20 = 0.225$$ **②**

b Make l the subject of the formula: $T = 2\pi\sqrt{\frac{l}{g}}$

Answer

You have to make the formula start with $l =$

so **first**, square both sides of the formula to get $T^2 = 4\pi^2 \frac{l}{g}$ **③**

Then multiply both sides by g to get: $gT^2 = 4\pi^2 l$ **③**

Last divide both sides by $4\pi^2$ to get: $\frac{gT^2}{4\pi^2} = l$ or $l = \frac{gT^2}{4\pi^2}$ **③**

Look out for

Like signs that multiplied or divided can be replaced with a **+** sign.

Unlike signs that multiplied or divided can be replaced with a **−** sign.

Whatever you **do to one side** of a formula you **must do to the other side** as well.

Key terms

Variable

Substitute

Formula

Equation

Exam-style questions

1 Work out the value of S when $u = -5$, $t = 10$ and $a = -4.9$

$S = ut + at^2$ **[2]**

2 Make t the subject of this formula.

$y = 5at^2 + 3s$ **[3]**

Exam tip

Remember to show the numbers when you substitute them into the formula.

CHECKED ANSWERS

Identities

REVISED

LOW

Rules

❶ A formula is an equation for working out the value of the subject of the formula.
❷ An expression is a collection of terms or variables that occur in formulae, equations and identities.
❸ An equation can be solved to find the value of an unknown variable.
❹ An identity is always **true** for all possible values of the variables.

Worked examples

a Here is a list of collections of terms.

i $5(2x - 3) = 10x - 15$
ii $5(2x - 3) = 8x + 2$
iii $12x^2y^3$
iv $p = 5(2x - 3)$

Write down the special mathematical name for each collection.

Answers

i This is an identity because the expression on both sides of the equals sign is identical ❹

ii This is an equation because the value of the variable has a unique value. The equation becomes $2x = 17$ or $x = 8.5$ ❸

iii This is an expression; x and y are variables and 12 is the coefficient of the expression ❷

iv This is a formula where p is the subject of the formula ❶

b Show that $\dfrac{3(x + 1)}{4} - \dfrac{2(x - 3)}{3} \equiv \dfrac{x + 33}{12}$.

Answer

b First, write the left-hand side over a common denominator of 12.

$$\frac{3 \times 3(x + 1) - 4 \times 2(x - 3)}{12}$$

Then expand the brackets, watching for the signs.

$$\frac{9x + 9 - 8x + 24}{12}$$

Collect like terms: $\dfrac{x + 33}{12}$ Which is identical to the right-hand side.

Look out for

If n is used to represent whole numbers then $2n$ is used to represent even numbers or multiples of **2** and $2n - 1$ or $2n + 1$ will then represent odd numbers.

Key terms

Variable

Term

Subject

Formula

Expression

Equation

Identity

Coefficient

Exam-style question

1 Find the value of p and q to make this expression into an identity.
$x^2 - 7x + 12 = (x + p)(x + q)$ **[2]**

Exam tip

Remember to always show each step in your working.

CHECKED ANSWERS

Using indices in Algebra

Rules

1 Index law for multiplying numbers or variables raised to a power is $a^n \times a^m = a^{n+m}$.

2 Index law for dividing numbers or variables raised to a power is $a^n \div a^m = a^{n-m}$.

3 Index law for raising a variable written as a power to a power is $(a^n)^m = a^{n \times m}$.

4 Index law for reciprocals or '1 over' is
$a^{-1} = \frac{1}{a}$ so $b^{-3} = \frac{1}{b^3}$ and $\frac{1}{c^{-1}} = c$.

5 Index law for roots is $\sqrt{x} = x^{\frac{1}{2}}$ so $\sqrt[3]{y} = y^{\frac{1}{3}}$ and $\sqrt[3]{r^4} = (\sqrt[3]{r})^4 = r^{\frac{4}{3}}$.

Key terms
Variable

Index

Power

Reciprocal

Root

Worked examples

a Simplify fully:

i $\dfrac{s^5 t^{\frac{3}{2}} u^{\frac{3}{5}}}{s^{\frac{5}{2}} t^3 u^{\frac{2}{5}}}$

ii $\dfrac{\sqrt[6]{x^4 y^{-3}}}{\sqrt[3]{x^{-4} y^3}}$

Look out for

Make sure that you can add, subtract, multiply and divide fractions and deal with positive and negative numbers.

Answers

2 4

i $s^{\left(5-\frac{5}{2}\right)} t^{\left(\frac{3}{2}-3\right)} u^{\left(\frac{3}{5}-\frac{2}{5}\right)} = s^{\frac{5}{2}} t^{-\frac{3}{2}} u^{\frac{1}{5}} = \dfrac{s^{\frac{5}{2}} u^{\frac{1}{5}}}{t^{\frac{3}{2}}}$ or $\dfrac{\sqrt{s^5} \times \sqrt[5]{u}}{\sqrt{t^3}}$ **5**

2 4 5

ii $x^{\left(\frac{4}{6}-\frac{4}{3}\right)} y^{\left(-\frac{3}{6}-\frac{3}{3}\right)} = x^{\left(\frac{2}{3}+\frac{4}{3}\right)} y^{\left(-\frac{1}{2}-1\right)} = x^{\frac{6}{3}} y^{\left(-\frac{3}{2}\right)} = \dfrac{x^2}{\sqrt{y^3}}$ **5**

b Find the value of n that makes this a true statement $x^n = \sqrt[3]{x^7 x^{-3}}$.

Answer

$\sqrt[3]{x^7 x^{-3}} = x^{\frac{7}{3}} \times x^{-\frac{3}{3}} = x^{\left(\frac{7}{3}+-\frac{3}{3}\right)} = x^{\left(\frac{7}{3}-\frac{3}{3}\right)} = x^{\frac{4}{3}}$ **5 1**

n is therefore equal to $\frac{4}{3}$.

Exam-style questions

1 Simplify fully:

a $\sqrt[4]{p^8 q^6 r^2}$ **[2]**

b $\dfrac{\sqrt[4]{x^2 y^{-3}}}{\sqrt[3]{x^{-4} y^3}}$ **[3]**

2 Find the value of n to make this a true statement.

$\left(e^{\frac{4}{3}}\right)^{-\frac{5}{n}} = \sqrt[4]{e^5}$ **[3]**

CHECKED ANSWERS

Exam tip

Remember to always show each step in your working so that you can gain marks.

Manipulating more expressions; algebraic fractions and equations

Rules

1 When you expand three pairs of brackets you multiply every term in the second bracket by every term in the third bracket then multiply the answer by each term in the first bracket.

2 To simplify an algebraic fraction you need to factorise the numerator and the denominator as a first step.

3 To solve an equation that has fractions in it you need to write every term in the equation as a fraction with the same common denominator.

Worked examples

a Expand and simplify $(x + 3)(2x - 4)(3x + 5)$.

Answer

1 $(x + 3)(2x - 4)(3x + 5)$

$2x \times 3x = 6x^2$ $-4 \times 3x = -12x$

$2x \times +5 = 10x$ $-4 \times + 5 = -20$

$= (x + 3)(6x^2 + 10x - 12x - 20)$

1 $= (x + 3)(6x^2 - 2x - 20)$

$x \times 6x^2 = 6x^3$ $3 \times 6x^2 = 18x^2$

$x \times -2x = -2x^2$ $3 \times - 2x = -6x$

$x \times -20 = -20x$ $3 \times -20 = -60$

$= 6x^3 - 2x^2 - 20x + 18x^2 - 6x - 60 = 6x^3 + 16x^2 - 26x - 60$

b i Simplify $\dfrac{4x^2 - 9}{2x^2 + 7x - 15}$ **ii** Solve $\dfrac{3x}{x+4} - \dfrac{2x}{x-3} = 1$

Key terms

Binomial

LCM

Numerator

Denominator

Cancelling

Look out for

Or you can use the table method as shown on page 23.

Look out for

Quadratic expressions that factorise when you have algebraic fractions.

Answers

i $\dfrac{(2x+3)(2x-3)}{(2x-3)(x+5)}$ **2** factorise

$\dfrac{(2x+3)(2x-3)}{(2x-3)(x+5)}$ cancel

$\dfrac{2x+3}{x+5}$

ii Write over the common denominator. **3**

$\dfrac{3x(x-3) - 2x(x+4)}{(x+4)(x-3)} = \dfrac{(x+4)(x-3)}{(x+4)(x-3)}$

Multiply both sides by $(x + 4)(x - 3)$ and collect like terms.

(cancel) $3x^2 - 9x - 2x^2 - 8x = x^2 - 3x + 4x - 12$

$x^2 - 17x = x^2 + x - 12$

$12 = 18x$ so $x = \dfrac{2}{3}$

Exam-style questions

1 a Write as a single fraction $\dfrac{3}{x+4} + \dfrac{5x}{x-4}$. **[3]**

   b Solve $\dfrac{2x}{x+5} - \dfrac{3}{x-4} = 2$. **[4]**

2 Prove that $(n + 1)^3 - (n + 1)^2 = n(n + 1)^2$ **[3]**

Exam tip

If you write down each step in your working you will make fewer mistakes and get the method marks.

CHECKED ANSWERS

Rearranging more formulae

Rules

1. You can replace the variables or letters in a formula with positive and negative numbers.
2. Use BIDMAS and rules for dealing with adding, subtracting, multiplying and dividing positive and negative numbers to find the value of the missing letter.
3. Use inverses to write the formula so that the required variable or letter is on its own on one side of the formula or equation.
4. If the subject of the formula appears twice then you will have to collect this variable on one side and factorise the variable outside a bracket.

Key terms

Variable

Subject

Formula

Equation

Inverse

Worked examples

a Here is a formula used in physics $\frac{1}{f} = \frac{1}{u} + \frac{1}{v}$.

Find the value of u when $f = 2$ and $v = 3$.

Answer

1. The first step is to substitute the values into the formula: $\frac{1}{2} = \frac{1}{u} + \frac{1}{3}$.

   Then get the value you want to find on one side of the equation

3. by subtracting the $\frac{1}{3}$ from each side: $\frac{1}{2} - \frac{1}{3} = \frac{1}{u}$ so $\frac{1}{6} = \frac{1}{u}$.

3. Then take the reciprocal of each side of the equation so $u = 6$.

b Make x the subject of this formula $y = \sqrt{\frac{x+k}{x-k}}$.

Look out for

The variable you have to make the subject of the formula appearing twice.

Answer

3. Firstly, square each side to get $y^2 = \frac{x+k}{x-k}$.

3. Then multiply both sides by $(x - k)$: $y^2(x-k) = x+k$.

2. Expand the bracket: $y^2x - y^2k = x + k$.

4. Collect the xs on one side: $y^2x - x = y^2k + k$.

4. Factorise the x and k outside a bracket: $x(y^2 - 1) = k(y^2 + 1)$.

3. Divide throughout by $(y^2 - 1)$: $x = \frac{k(y^2+1)}{y^2-1}$.

Exam-style questions

1. Make m the subject of the formula $P = c(8 - 3m) + 2m$. **[3]**

2. Make T the subject of the formula $K = \sqrt{\frac{PT}{S+T}}$. **[3]**

Exam tip

If you write down each step in your working you will make fewer mistakes.

CHECKED ANSWERS

Special sequences

Rules

1. The difference between each term in a triangle number sequence goes up by one extra each time: 1, 3, 6, 10, 15, 21,...
2. The difference between each term in a square number sequence goes up by the same extra number each time: 1, 4, 9, 16, 25, 36, ...
3. The difference between each term in the Fibonacci sequence is also the Fibonacci sequence: 1, 1, 2, 3, 5, 8, 13, 21 ...
4. By putting whole numbers (1, 2, 3, ...) into the nth term you can build up each term in the sequence.

Worked examples

a   Write down the first five terms of the sequences with nth terms
  i $n(n + 1)$
  ii $3n^2 + 1$

Answers
Put the values 1, 2, 3, 4, 5 instead of n to calculate the first 5 terms. ❹
  i $1 \times 2, 2 \times 3, 3 \times 4, 4 \times 5, 5 \times 6$ gives 2, 6, 12, 20, 30
  ii $3 \times 1^2 + 1, 3 \times 2^2 + 1, 3 \times 3^2 + 1, 3 \times 4^2 + 1, 3 \times 5^2 + 1$, gives
      $3 \times 1 + 1, 3 \times 4 + 1, 3 \times 9 + 1, 3 \times 16 + 1, 3 \times 25 + 1$,
      or 4, 13, 28, 49, 76

b   Find the nth terms of these number patterns.
  i   2, 5, 10, 17, 26, ...
  ii 4, 12, 24, 40, 60, ...

Answers
  i   The difference between each term is 3, 5, 7, 9; the difference goes up in two's so it must be a square number pattern. ❷
      The nth term is $n^2 + 1$, one more than the square numbers.
  ii The difference between each term is 8, 12, 16, 20; the difference goes up in 4 extra each time so it must be a triangular number pattern. ❶
      The nth term is $2n(n + 1)$, four times the triangle numbers.

Look out for

The nth term for square number is n^2.

The nth term for triangle number is $\frac{n(n+1)}{2}$.

Key terms

Sequence

Triangle numbers

Square numbers

Fibonacci numbers

Term

nth term

Difference

Exam-style questions

1 Rachel makes a pattern from squares.

  a   Find the nth term of Rachel's pattern. **[2]**
  b   How many squares are there in pattern number 20? **[1]**

2 Here is a number pattern   6, 12, 20, 30, 42, ...

   Find the nth term of the pattern. **[3]**

Exam tip

Always look for the difference between each term in a number pattern to help decide which type of pattern it is.

CHECKED ANSWERS

Quadratic sequences

Rules

1. In a quadratic sequence the difference between the terms increases by the same number each time.
2. In a quadratic sequence the difference between the differences is always the same number. This is called the second difference.

Sequence		2		8		18		32		50		72
First difference			6		10		14		18		22	
Second difference				4		4		4		4		

3. By putting whole numbers (1, 2, 3, …) into the nth term you can build up the sequence.

Worked examples

a The nth term of a quadratic sequence is: $2n^2 - 1$. The mth term of a different quadratic sequence is: $98 - (m + 1)^2$. Which numbers are in both sequences?

Answer

List both sequences by putting in values of 1, 2, 3, etc. for n.

$2n^2 - 1$ gives: $2 \times 1^2 - 1, 2 \times 2^2 - 1, 2 \times 3^2 - 1$, etc. ③

$98 - (m + 1)^2$ gives: $98 - (1 + 1)^2, 98 - (2 + 1)^2, 98 - (3 + 1)^2$, etc. ③

Sequences are: 1, 7, 17, 31, 49, 71, 97, …

and: 94, 89, 82, 73, 62, 49, 34, 17, –2, …

So 17 and 49 are in both sequences.

b Find the nth term of this sequence: 3, 7, 13, 21, 31, 43, …

Answer

Sequence		3		7		13		21		31		43
① First difference			4		6		8		10		12	
② Second difference				2		2		2		2		

As the second difference is always 2, the sequence is quadratic.

The sequence builds by $1 \times 2 + 1, 2 \times 3 + 1, 3 \times 4 + 1, 4 \times 5 + 1$, etc.

The nth term is: $n(n + 1) + 1$

Look out for

If the second difference is 2 then the coefficient of n^2 is always 1.

Watch for triangular numbers in a quadratic sequence.

Key terms

Quadratic

Sequence

Difference

Second difference

Exam-style questions

1 The nth term of a quadratic sequence is:
$(n + 1)^2 - 2$

The mth term of a different quadratic sequence is: $50 - m^2$

Which numbers are in both sequences? **[3]**

2 Find the nth term of this sequence:
7, 13, 23, 37, 55, … **[3]**

Exam tips

Look out for square numbers in the sequences.

If the second difference is 2, start with n^2.

If the second difference is 4, start with $2n^2$.

If the second difference is 6, start with $3n^2$.

*n*th term of a quadratic sequence

Rules

1 In a quadratic sequence the difference between the terms increases by the same number each time.
2 In a quadratic sequence the difference between the differences is always the same number. This is called the second difference.

Sequence	2		8		18		32		50		72
First difference		6		10		14		18		22		
Second difference			4		4		4		4			

3 The coefficient of the n^2 term is always half the second difference.
4 The general form of the *n*th term of a quadratic sequence is $an^2 + bn + c$.
5 Put in values of $n = 1$ and $n = 2$ to find the values of b and c.

Worked examples

a Find the *n*th term of this quadratic sequence: 4   9   18   31   48 ...

Answer

First you write down the sequence   4 9 18 31 48

1 Then you work out the 1st difference   5 9 13 17

2 Then the 2nd difference   4 4 4

3 The 2nd are all 4 so the coefficient of n^2 is half of 4 which is 2.

4 The general form of the sequence becomes $2n^2 + bn + c$.

Now we need to find b and c by putting in values for n.

5 When $n = 1$ $2 \times 1^2 + b + c = 4$ so $2 + b + c = 4$ or $b + c = 2$

When $n = 2$ $2 \times 2^2 + 2b + c = 9$ so $8 + 2b + c = 9$ or $2b + c = 1$

Subtracting the two blue equations gives $b = -1$.

This means we now have $2n^2 - n + c$ for the *n*th term.

Using $n = 1$ again we get $2 \times 1^2 - 1 + c = 4$ so $2 - 1 + c = 4$ or $c = 3$.

The *n*th term is $2n^2 - n + 3$.

> **Look out for**
>
> Difference and Second difference being the same.

> **Key terms**
>
> Difference
>
> Second difference
>
> Quadratic sequence
>
> Coefficient
>
> General form of a quadratic sequence

b The first and third terms of the quadratic sequence $n^2 + bn + c$ are 5 and 13. Work out the first five terms of the sequence.

Answer

4 General term is $an^2 + bn + c$ so in this case $a = 1$.

5 When $n = 1$ $1^2 + b + c = 5$ so $1 + b + c = 5$ or $b + c = 4$

When $n = 3$ $3^2 + 3b + c = 13$ so $9 + 3b + c = 13$ or $3b + c = 4$

Subtracting blue equations gives $2b = 0$ so $b = 0$ and therefore $c = 4$.

*n*th term is $n^2 + 4$ so first five terms are 5, 9, 13, 20 and 29.

Exam-style questions

1 Find the *n*th term of this quadratic sequence
   2   9   22   41   66 ... **[5]**
2 The *n*th term of a quadratic sequence is $2n^2 - 3n + 6$.
   The *n*th term of a different quadratic sequence is $(2n - 1)(n + 4)$.
   One number has the same position in both sequences.
   Find the number. **[4]**

> **Exam tip**
>
> It is always a good idea to check your *n*th term to see if it gives the original sequence.

CHECKED ANSWERS

The equation of a straight line

Rules

❶ Vertical lines which are parallel to the y-axis have the equation $x =$ a number.

❷ Horizontal lines which are parallel to the x-axis have the equation $y =$ a number.

❸ Slanting lines have the equation $y = mx + c$

❹ To find m, the gradient of a slanting line, find the coordinates of two points on the line and divide the difference of their y-coordinates by the difference of their x-coordinates.

❺ The value of c is the y-coordinate of the point where the line crosses the y-axis.

Look out for

Draw a diagram to help answer the question.

Parallel lines have the same gradient.

Lines with a positive gradient go from bottom left to top right.

Lines with a negative gradient go from top left to bottom right.

Worked example

a Here is a straight line drawn on a coordinate grid. Find the equation of the line.

Answer
Two points on the line are (–1, 2) and (1, 6) ❸

The gradient of the line is found by finding the lengths of the horizontal and vertical lines and dividing them.

The vertical length is 6 – 2 = 4 units and the horizontal length 1 – –1 = 2 units

Divide the vertical length by the horizontal length.

The gradient is 4 ÷ 2 = 2 ❹

The intercept, c, on the y-axis is at 4 ❺

The gradient is positive as it goes from bottom left to top right.

Therefore the equation of the line is $y = 2x + 4$.

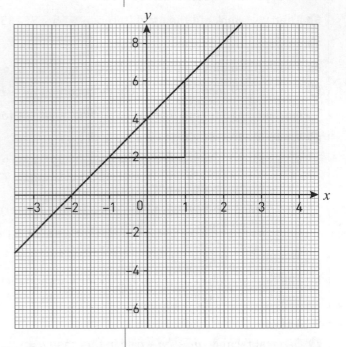

b Here are some straight lines. Which of them are parallel?

P $y = 3x + 2$ **Q** $y + 3x = 2$
R $3y = 3 – 9x$ **S** $9x + 3y = 12$
T $3x – y = 5$

Answer
Firstly write all the equations in the form $y = mx + c$

P $y = 3x + 2$ **Q** $y = –3x + 2$
R $y = –3x + 1$ **S** $y = –3x + 4$ **T** $y = 3x – 5$

Now check for the ones that have the same gradient. **P** and **T** have the same gradient (3) and also **Q**, **R** and **S** have the same (–3).

Key terms

Gradient

Intercept

Parallel

Exam-style questions

1 Write down an equation of a line that is parallel to $y = 2x + 3$ that passes through the point (0, –1) **[2]**

2 Here are some straight lines.
   P $y = 2x + 3$ **Q** $y + 2x = 1$
   R $2y = 3 – 4x$ **S** $8x – 4y = 12$ **T** $6x – 3y = 15$
   Which of them are parallel? **[3]**

Exam tip

Always write your equations in the form $y = mx + c$

Plotting quadratic and cubic graphs

Rules

1. Always draw a table of values to help plot the points on the grid.
2. Start with the value 0 and put in positive values first.
3. Plot the points and join them with a smooth curved line.
4. You can use the graph to read off values from one axis to the other.
5. A quadratic graph will be in the shape of a letter U or an ∩.
6. A cubic graph will be in the shape of a letter ∽.

Worked example

a Here is the graph of $y = x^2 - 3x - 2$ for values of x from −2 to +4. ③ ④ ⑤

  i What is the minimum value of $x^2 - 3x - 2$?

  ii For what values of x is y negative?

Answers

i The minimum value is at the bottom of the U. The minimum value is when x is 1.5 so $y = -4.25$

ii y is negative when it goes below the x-axis so between $x = -0.6$ and 3.6

b **i** Draw the graph of $y = x^3 - 5x + 2$ for values of x from −3 to +3 ③ ④ ⑥

  ii Solve the equation $x^3 - 5x + 2 = 0$

Answers

i ① First you make a table of values. ② Start with 0 which means that $y = +2$. Then work out the positive values of x first. Finally do the negative values.

x	−3	−2	−1	0	1	2	3
x^3	−27	−8	−1	0	1	8	27
$-5x$	15	10	5	0	−5	−10	−15
+2	2	2	2	2	2	2	2
$y =$	−10	4	6	2	−2	0	14

ii The solutions to the equation are where the curve cuts the x-axis at $x = -2.4$ or at $x = 0.4$ or at $x = 2$

Look out for

Your y values in the table should be symmetrical for a quadratic graph.

Make sure you join the points with a smooth curve.

If you have to solve an equation then a quadratic will have 2 answers and a cubic will have 3 answers, some may be the same.

Key terms

Quadratic

Cubic

Equation

Maximum

Minimum

Exam-style questions

1. **a** Draw the graph of $y = x^2 - 4x + 3$ for values of x from −2 to +4 **[4]**
   b Use your graph to solve the equation $x^2 - 4x + 3 = 0$ **[2]**
2. Draw the graph of $y = x^3 - 5x + 2$ for values of x from −3 to +3 **[4]**

Exam tip

Always look for solving an equation as the last part to a graph question.

Finding equations of straight lines

Rules

1. Slanting lines have the equation $y = mx + c$
2. m is the gradient of the line and measures how steep the line is.
3. To find m, the gradient of a slanting line, find the co-ordinates of two points on the line and divide the difference of their y-co-ordinates by the difference of their x-co-ordinates.
4. The value of c, the intercept on the y-axis, is the y-co-ordinate of the point where the line crosses the y-axis.

Worked examples

a **L** and **M** are two straight lines. **L** has a gradient of 3 and crosses the y-axis at $(0, 4)$. **M** has a gradient of -2 and passes through the point $(-1, 5)$. Find the equations of the two lines.

Answer

The equation of **L** is $y = 3x + 4$. ❶

3 is the gradient of the line so is the value of m. ❷

$(0, 4)$ is on the y-axis so the value of c is 4 ❹

The equation of **M** is $y = -2x + c$ ❶

-2 is the gradient of the line so the value of m is -2 ❷

As we do not know the value of c you need to put the co-ordinates $(-1, 5)$ into the equation so with $x = -1$ and $y = 5$ you get $5 = -2 \times -1 + c$ or $5 = +2 + c$ so $c = 3$ ❹

The equation of **M** is $y = -2x + 3$

b **P** and **Q** are two straight lines. **P** is parallel to the line $2x + y = 5$ and passes through $(2, 9)$. **Q** passes through the points $(-2, 7)$ and $(4, -5)$. Find the equations of the two lines.

Answer

P is parallel to $2x + y = 5$. This can be written as $y = -2x + 5$ so -2 is the gradient of the line so the value of m is -2 ❷. The equation of **P** is therefore $y = -2x + c$ ❶. To find the value of c use a co-ordinate that lies on the line, $(2, 9)$, put this into the equation so with $x = 2$ and $y = 9$ you get $9 = -2 \times 2 + c$ or $9 = -4 + c$ so $c = 13$ ❹.

The equation of **P** is $y = -2x + 13$. Always sketch the line so you can check the gradient.

Gradient is $(7 - -5) \div (-2 - 4) = 12 \div -6 = -2$ ❸

The equation of **Q** is $y = -2x + c$ ❶

Substitute a point into the equation e.g. $7 = -2 \times -2 + c$ so $c = 7 - 4 = 3$ ❹

The equation of **Q** is $y = -2x + 3$

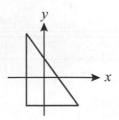

Look out for

Parallel lines have the same gradient.

Lines with a positive gradient go from bottom left to top right.

Lines with a negative gradient go from top left to bottom right.

Key terms

Gradient

Intercept

Equation

Parallel

Exam tip

If given the co-ordinates, always sketch a diagram so you can check whether the gradient of the line is positive or negative.

Exam-style questions

1. Find the equation of the straight line that is parallel to $y = 3x + 2$ and passes through $(1, 6)$ **[3]**

2. Find the equation of the straight line that passes through the points $(-2, 5)$ and $(3, -5)$ **[3]**

CHECKED ANSWERS

Polynomial and reciprocal functions

Rules

1. Straight line graphs are in the form $y = mx + c$
2. Quadratic graphs are in the form $y = ax^2 + bx + c$ and are in the shapes of a U or ∩.
3. Cubic graphs are in the form $y = ax^3 + bx^2 + cx + d$ and are in the shape of an ∽.
4. Reciprocal graphs are in the form $y = \frac{k}{x}$ and have two parts. They approach but never touch two straight lines; the straight lines are called asymptotes.

Worked example

Here is the graph of $y = 6 - x^2$

i On the same grid, draw the graph of $y = \frac{1}{2x}$

ii Write down the equations of the asymptotes.

Answers

i First work out a table of values. ❹

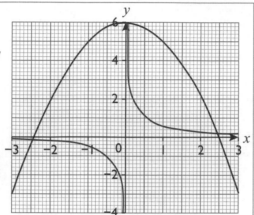

x	−3	−2	−1	0	1	2	3
$y = \frac{1}{2x}$	−0.17	−0.25	−0.5	No value	0.5	0.25	0.17

As there is no value for $\frac{1}{2x}$ when $x = 0$ you will need to use values of x between

$-1 \leqslant x < 0$ and $0 < x \leqslant 1$

x	−0.5	−0.2	−0.1	0	0.1	0.2	0.5
$y = \frac{1}{2x}$	−1	−2.5	−5	No value	5	2.5	1

ii The asymptotes are $y = 0$ (the x-axis) and $x = 0$ (the y-axis). ❹

Look out for

When you have reciprocal graphs there will always be one value that you cannot calculate, normally when $x = 0$

Key terms

Plot

Sketch

Straight line

Quadratic

Cubic

Reciprocal

Asymptote

Exam-style questions

1 a On the same grid sketch the graphs of $y = x$, $y = x^2$ and $y = x^3$ **[3]**
  b Which points lie on all three lines? **[2]**

2 Jill's car fuel consumption, f, changes as her speed, s, increases.

The fuel consumption is given by the formula $f = 60 - \frac{60}{s}$

What value is the fuel consumption approaching as she increases her speed?

You must explain your answer. **[4]**

Exam tips

Draw or plot means a graph needs to be accurate.

Sketch means you need show the main features of the graph.

Perpendicular lines

MEDIUM

Rules

❶ If a straight line, **L**, has a gradient of m then the gradient of a straight line perpendicular to **L** has a gradient of $-\frac{1}{m}$.

❷ If two lines with gradients m_1 and m_2 are perpendicular then $m_1 \times m_2 = -1$.

Worked example

a A straight line, **L**, is perpendicular to the line $y + 2x = 3$. It passes through the point $(6, 2)$.

Find the equation of **L**.

Answer

Firstly arrange the equation into the form $y = mx + c$.

So $y + 2x = 3$ becomes $y = -2x + 3$.

❶ This means the gradient is -2.

❷ The gradient of **L** is therefore $\frac{1}{2}$ because $-2 \times \frac{1}{2} = -1$.

The equation of **L** is therefore $y = \frac{1}{2}x + c$ so use the point $(6, 2)$ to find c.

$2 = \frac{1}{2} \times 6 + c$ means that $c = -1$ so **L** has an equation $y = \frac{1}{2}x - 1$ or $2y = x - 2$.

b A straight line passes through P, at $(5, 0)$ and Q, at $(0, 12)$.

Explain why the straight line that passes through P that is perpendicular to PQ also passes through $(17, 5)$.

Answer

❶ & ❷ The gradient of PQ is $-\frac{12}{5}$ so the gradient of the perpendicular line is $\frac{5}{12}$.

The equation is therefore $y = \frac{5}{12}x + c$ so at $(5, 0)$ it is $0 = \frac{5}{12} \times 5 + c$.

Which means $c = -\frac{25}{12}$. The equation is $y = \frac{5}{12}x - \frac{25}{12}$ or $12y = 5x - 25$.

At $(17, 5)$ $12 \times 5 = 60$ and $5 \times 17 - 25$ also $= 60$.

This means the line $12y = 5x - 25$ passes through $(17, 5)$.

Key terms

Perpendicular lines

Positive and negative gradient

Look out for

$y = mx + c$ and $y = -\frac{1}{m}x + c$ are perpendicular lines $m_1 \times m_2 = -1$

Exam tip

It is always a good idea to make a sketch of the problem to help you know what is going on.

Exam-style questions

1 Find the equation of the straight line that passes through $(2, 1)$ and is perpendicular to $y = 4x - 3$. **[3]**

2 **L**, **M** and **N** are three straight lines.

  L has equation $y + 2x = 5$

  M is perpendicular to **L** and passes through $(2, 1)$

  N has equation $y = 3$

  Find the area of the triangle formed by the three straight lines. **[4]**

CHECKED ANSWERS

Exponential functions

Rules

1. An exponential growth function has the form $f(x) = ab^x$ if x is positive.
2. An exponential decay function has the form $f(x) = ab^x$ if x is negative.
3. The general form of an exponential function is $f(x) = ab^{kx}$.
4. When $x = 0$ the intercept on the vertical axis $(x = 0)$ gives the value of a.
5. The value of b is the number $f(x)$ is multiplied by as x increases.
6. The value of k is the number x is multiplied to make $kx = 1$.

Worked example

a Here is a graph of the population of rabbits in an enclosure. It shows the population each month.

   i Explain why the population growth is exponential.

   ii Find the equation of the graph.

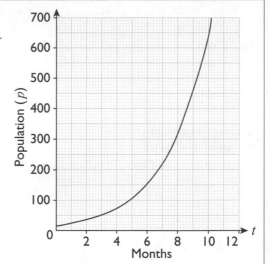

Answers

i When $t = 0$, $P = 20$

  $t = 2$, $P = 40$ $t = 4$, $P = 80$

  $t = 6$, $P = 160$ $t = 8$, $P = 320$

This means the population doubles every 2 months.

This is a typical exponential function of the form $f(x) = ab^{kx}$ as the rate of increase is increasing each month. ❸

ii This equation will be of the form $P = ab^{kt}$.

This means that x, or in this case t, will be positive. ❶

When $t = 0$, $P = 20$ which means $a = 20$. ❹

$b = 2$ as the population doubles ❺

$k = \frac{1}{2}$ as $2 \times \frac{1}{2} = 1$ ❻

The equation of the graph will be $P = 20 \times 2^{\frac{t}{2}}$.

Look out for

Population growth and compound interest are examples where rule ❶ works.

Population decay and depreciation are examples where rule ❷ works.

Key terms

Function

Exponential

Population

Growth and decay

Compound Interest

Depreciation

Exam-style questions

1 Here are the population figures for a troop of monkeys in a forest.

t (years)	0	1	2	3	4	5	6
P	1600	1270	1008	800	635	504	400

  a Explain why the population decrease is exponential. **[2]**
  b Find the equation of the graph. **[3]**

Exam tip

Always look for simple relationships between some of the data you are given.

Trigonometric functions

Rules

1. The general form of a trig function is $f(x) = A \sin(Bx + C) + D$ where sin could be replaced by any of the trigonometrical operators, for example, sin or cos or tan.
2. Changing the value of A changes the height of the graph.
3. Changing the value of B changes how quickly the cycle repeats itself.
4. Changing the value of C translates the graph parallel to the x-axis.
   If C is positive the graph moves left and if C is negative it moves to the right.
5. Changing the value of D translates the graph parallel to the $f(x)$ or y-axis
   If D is positive the graph moves up and if D is negative it moves down.

Key terms

Trigonometric functions

Periodic

Cyclical

Worked example

a Here is a graph of a trig function $f(x)$.

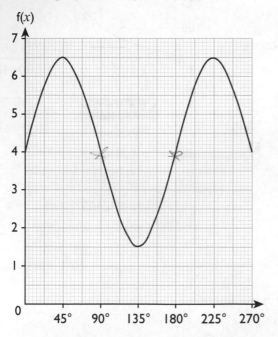

i Work out the equation of the graph.

ii Use the graph or otherwise to solve $f(x) = 3$ where $0° \leqslant x \leqslant 360°$.

Answers

i Looking at the graph you can see it has the form of the sine / cosine function.

5. It has been moved up from the x-axis by 4 units so the value of D is 4.

4. For the sine function the curve starts on the central horizontal line or the x-axis when $x = 0$ so $C = 0$.

4. For the cosine function the curve would start at a maximum point so it has moved 45° to the right so in this case $C = -45°$.

Look out for

Trigonometric functions are always periodic.

Trigonometric functions always repeat themselves.

Trigonometric functions are always symmetrical.

This diagram shows the angles when trig ratios are positive.

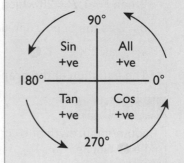

❸ A normal sine or cosine curve will repeat itself every 360°. In this case the curve repeats every 180° so the value of B will be 2 as 360 ÷ 180 = 2.

❷ The height / amplitude of the sine curve is normally 1. In this case it is 2.5 so $A = 2.5$.

❶ The equation is f(x) or $y = 2.5 \times \sin 2x + 4$ or f(x) or $y = 2.5 \times \cos 2(x - 45°) + 4$.

ii There should be four solutions to the equation f(x) = 3 though only two are shown in the diagram. These are $x = 104°$ and 166° which is 135° ± 31°.

From the symmetry of the graph the other values are 315° ± 31° so $x = 284°$ and 346°.

Exam-style questions

1 Solve the equation $4 \cos 3x = 2$ for values of x between 0° and 360°. **[4]**

2 The height, h metres, of water in a harbour during one day is modelled by the formula $h = 4 \sin 30(t - 3) + 5$ for values of t from $0 \leqslant t \leqslant 24$.
   a What is the value of t when the value of h is a maximum? **[2]**
   b What is the value of t when the value of h is a minimum? **[2]**
   c Find the value of t when the height of water in the harbour is 5 metres. **[4]**

Exam tips

Always make sure you give all the solutions to a trig equation.

CHECKED ANSWERS

Mixed exam-style questions

1 Bobbi uses this formula to work out the time, t minutes, it takes to cook a chicken of weight w kg.
   $t = 40w + 20$
   Bobbi wants a chicken weighing 2 kg to be cooked at 12 noon.
   At what time should she put the chicken into the oven? **[3]**

2 A square has a perimeter of $(40x + 60)$ cm.
   A regular pentagon has the same perimeter as the square.
   Show that the difference between the length of the sides of the two shapes is $(2x + 3)$ cm. **[3]**

3 The orange square is formed by cutting the 4 blue right-angled
   triangles each with base of length $x + 2$ and height of length $5x - 3$,
   from each corner as shown.
   Show that the area of the orange square is $13(2x^2 - 2x + 1)$ **[5]**

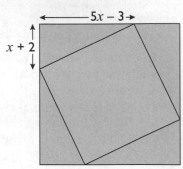

4 Here is a T shape drawn on part of a 10 by 10 grid.

1	2	3	4	5	6
11	12	13	14	15	16
21	22	23	24	25	26
31	32	33	34	35	36
41	42	43	44	45	46

   The shaded T is called T_2 because 2 is the smallest number in the T. T_2 is the sum of all the
   numbers in the T shape; so $T_2 = 45$.
   a Find an expression, in terms of n, for T_n. **[3]**
   b Explain why T_n cannot equal 130. **[2]**

5 Find the nth term of this quadratic sequence.
   1   4   11   22   37 ... **[5]**

6 a Make x the subject of the formula $y = 2\pi\sqrt{\frac{3x+5}{x}}$. **[3]**
   b Find the LCM and HCF of $12a^3b^2c^3$, $18a^2b^3c^4$ and $24a^3b^2c$. **[3]**

7 a Simplify $\frac{x^2-16}{2x^2-3x-20}$. **[3]**

8 Find the value of n to make this a true statement: $(x^{-n})^5 = \frac{\sqrt{x^3}}{\sqrt{x^n}}$. **[3]**

9 Here is a sequence of numbers: $2, 1, \frac{1}{2}, \ldots$
   a Find the seventh term in the sequence. **[2]**
   b For what values of n will the nth term be less than 0.001? **[3]**

10 The cost of hiring a car from **Cars 4 U** is £20 plus
a daily rate.
  a Work out the daily rate. **[2]**
Sid wants to compare the cost of hiring a car from
Cars 4 U and from **Car Co** who charge £25 for each
day of hire.
Sid hires cars for different periods of time. He wants to
use the cheaper company.
  b Which of these two companies is the cheaper to hire
    the car from?
You must show your working and explain
your answer. **[3]**

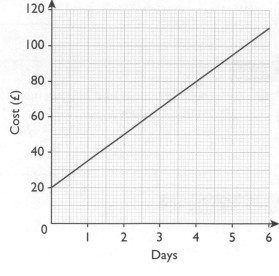

11 The line *l* has a gradient of 2 and passes through $(2, -3)$.
  a Find the equation of the straight line *l*. **[2]**
  b Explain whether the point $(3, -1)$ lies on the line *l*. **[2]**
  c Find the equation of the straight line perpendicular to
    l that also passes through $(2, -3)$. **[3]**

12 Here is the graph of $y = x^2 - 4x + 3$.
  a Write down the minimum value of *y*. **[1]**
  b Find the points where the line $x + y = 4$
    crosses the curve. **[2]**

13 A quadratic function passes through the points $(2, 0)$ and $(0, 4)$.
The function has only got one root.
  a Sketch the graph of the function. **[2]**
  b Find an equation of the function. **[2]**

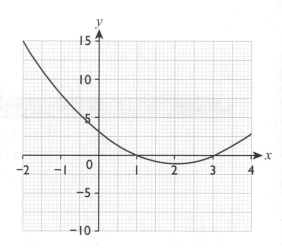

14 Alison is *x* years old. Alison is 2 years older than
Bethany. Cathy is twice as old as Bethany.
The total of their ages is 50.
What is Cathy's age? **[4]**

15 Find graphically the vertices of the triangle formed by
the straight lines with equations:
$x + y = 5;$ $y = 2x + 3;$ $2y = x - 3$ **[4]**

16 A rectangle has an area of $x^2 - 12x + 32$.
Find a possible algebraic expression for the
perimeter of this rectangle. **[3]**

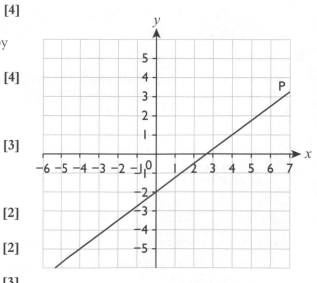

17 The straight line **P** has been drawn on the
co-ordinate grid.
  a Find an equation of the line **P**. **[2]**
  b Find an equation of the line parallel to **P** passing
    through $(-3, -1)$. **[2]**
  c Find an equation of the line perpendicular to
    the line **P** that passes through $(-1, 1)$. **[3]**

18 Here is the graph that shows the depth of water in
a harbour one day.
A ship is allowed to enter and leave the harbour
between 06:00 and 18:00. It needs a 6 metre depth of
water in the harbour for the large ship to enter.
Between what times can the ship enter the harbour
on this day? **[2]**

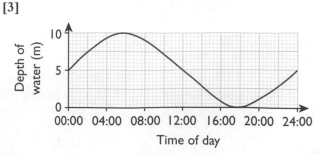

Trial and improvement

MEDIUM

Rules

❶ When solving by trial and improvement, always look to check your answer correct to one more decimal place than asked in the question.

For example, if solving to find a solution correct to 2 decimal places and your answer is between 5.47 and 5.48, do not look at which seems closer, you must check, a trial at 5.475 is needed to confirm where the solution lies.

Worked example

a $x^2 - 3x = 12$ has a solution between 5 and 6.

Find this solution correct to 1 decimal place.

Answer

First check the working using $x = 5$ and $x = 6$, then try half way between, $x = 5.5$.

By then you will know which way to check next.

Here $x = 5.5$ is still too large, so try a smaller value of x.

x	$x^2 - 3x$	
5	25 − 15 = 10	too small
6	36 − 18 = 18	too large
5.5	30.25 − 16.5 = 13.75	too large
5.4	29.16 − 16.2 = 12.96	still too large
5.3	28.09 − 15.9 = 12.19	still too large
5.2	27.04 − 15.6 = 11.44	too small

So far we find that the solution must be between 5.2 and 5.3, as one of these solutions is too large and the other too small.

❶ Now an important check.

We need to look at 5.25, half way between the solutions where it changes from too large to too small.

5.3	28.09 − 15.9 = 12.19	still too large
5.2	27.04 − 15.6 = 11.44	too small
5.25	27.5625 − 15.75 = 11.8125	too small

Now we know:

5.2	5.25	↑	5.3
too small	too small		too large

The solution must lie between these two values, so to 1 decimal place it must be **5.3**.

Key terms

Decimal place

Root

Significant figures

Solution

Solve

Exam tip

You have to look at 3 decimal places if the question says 'correct to 2 decimal places', but remember to give your answer then to the 2 decimal places asked for in the question.

Exam-style questions

1 The equation $3x^3 + x^2 = 54$ has a solution between 2 and 3.

Find this solution correct to 2 decimal places.

CHECKED ANSWERS

Linear inequalities

Rules

1. When you solve an inequality you need to keep the sign pointing the same way.
2. Changing the direction of the inequality sign is the same as multiplying by –1 so if you swap sides in an inequality you swap signs. So if $-5 > x$ then $-x > 5$ so $x < -5$.
3. Use the same techniques to solve an inequality as you do to solve an equation.
4. A filled-in circle on a number line means the inequality is $\geqslant$ or $\leqslant$. ●
5. An open circle on a number line means the inequality is $<$ or $>$. ○
6. Always define your variables when you solve an inequality problem.

Worked examples

a i Write down the inequality shown on this number line.

Answer

The left hand end of the inequality is at –2 and the right hand end is at 3. ❹ ❺ The circle at –2 is open and the circle at 3 is filled in so the inequality is: $-2 < x \leqslant 3$. ❹ ❺

ii Solve the inequality $5x + 3 > 7x - 4$

Answer

Keep the xs on the side that has the most of them. ❶
So $-5x$ from each side: $3 > 2x - 4$ ❸
Now add 4 to each side: $7 > 2x$
Divide by 2 gives: $3.5 > x$
Swap sides and change sign: $x < 3.5$ ❷

b Amy is 3 years older than Beth. Ceri is twice as old as Beth. The total of their ages is less than 39.
Show that Ami must be less than 12 years old.

Answer

First define your variable so let Ami's age be x ❻. Then write down the other ages.

Beth will be 3 years less than Ami: $x - 3$

Ceri's age will be twice Beth's age: $2(x - 3)$

Then set up the inequality: $x + x - 3 + 2(x - 3) < 39$

Then solve it: $2x - 3 + 2x - 6 < 39$ ❸. $4x - 9 < 39$

so $4x < 48$ so $x < 12$ and therefore Ami is less than 12 years old.

Look out for

Always use the same techniques for solving inequalities as you do to solve equations.

Key terms

$<$ less than

$>$ greater than

$\leqslant$ less than or equal to

$\geqslant$ greater than or equal to

Exam-style questions

1 a On a number line write down the inequality $-3 \leqslant x < 4$ **[2]**
  b Solve the inequality $3(2y - 4) \leqslant 6$ **[3]**

2 Bobbi thinks of a whole number, she adds 10 to it and then divides by 5. The answer is less than 4.
What numbers could Bobbi have thought of? **[4]**

Exam tip

Always check the answer to inequalities by substituting your answer back into the question.

CHECKED ANSWERS

Solving simultaneous equations by elimination and substitution

Rules

❶ If the coefficients of both of the variables are different then you must multiply the equations by a number so that the coefficients of one variable are the same.
❷ To eliminate the variable if the coefficients have the same sign you subtract the two equations; if the signs are different then you add the two equations.
❸ Once you have found the value of the other variable you substitute it into one of the original equations to find the eliminated variable.
❹ If the coefficient of one of the variables is 1 then rearrange that equation so that it becomes the subject e.g. $x =$ or $y =$
❺ Substitute the rearranged equation into the second equation.
❻ Solve the new equation for one variable.
❼ Substitute this variable into the first equation to find the other variable.

Worked examples

a Solve $3x + 4y = 2$ eqn. 1, $4x - 5y = 13$ eqn. 2

Answer

The coefficients of the variables x and y are not the same so you need to multiply each equation by a number to make them the same.

You could multiply eqn. 1 by 4 and eqn. 2 by 3 so that the x coeff. is 12

You could multiply eqn. 1 by 5 and eqn. 2 by 4 so that the y coeff. is 20

It is easier to add than subtract so we eliminate the y

$3x + 4y = 2$ eqn. 1 × 5 gives $15x + 20y = 10$ ❶
$4x - 5y = 13$ eqn. 2 × 4 gives $16x - 20y = 52 +$ ❶
Adding the two equations gives $31x \quad = 62$ **so $x = 2$** ❷

Substituting $x = 2$ into eqn. 1 gives ❸ $6 + 4y = 2$, so $y = -1$

b Solve $2x + y = 3$ eqn. 1 $3x - 4y = 10$ eqn. 2

Answer

The coefficient of y in eqn. 1 is equal to 1 so $y = 3 - 2x$ ❹

We substitute $y = 3 - 2x$ into eqn. 2 to get $3x - 4(3 - 2x) = 10$ ❺

Multiply out the bracket to get $3x - 12 + 8x = 10$ ❻

Simplify the left-hand side to get $11x - 12 = 10$ ❻

So $11x = 22$ so $x = 2$ ❻

Substituting $x = 2$ into $y = 3 - 2x$ gives $y = 3 - 4$, so $y = -1$

Look out for

The coefficients of the variables need to be made the same for the elimination method.

If one of the variables has a coefficient of 1 use the substitution method.

Key terms

Simultaneous equations
Coefficient
Variable
Subject
Substitute
Eliminate
Solve

Exam tip

Always check your answer by substituting back into the original equations.

Exam-style questions

1 Solve $2a + 3b = 13$
     $5a - 2b = 4$ **[3]**

2 A coach company has s superior coaches and d ordinary coaches. The company has four times as many ordinary coaches as superior coaches. An ordinary coach has 50 seats and a superior coach has 25 seats. The coach company has a total of 675 seats available. How many coaches of each type does the company have? **[4]**

CHECKED ANSWERS

Using graphs to solve simultaneous equations

Rules

1. Form an equation for each part of your problem in the form $y = mx + c$
2. On a co-ordinate grid draw the two lines so that they intersect.
3. The solution of the simultaneous equations are the co-ordinates of the point of intersection.

Worked examples

a G-gas and P-gas sell gas. G-gas charges £10 a month and 20p a unit. P-gas charges £20 a month and 10p a unit. Which company is cheaper?

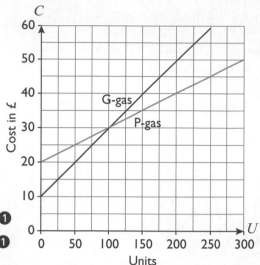

Answer

Firstly set up equations.

$C = 0.2u + 10$ for G-gas. ❶

$C = 0.1u + 20$ for P-gas. ❶

Use the same units (£).

Then draw the graphs. ❷

Use the intercept and gradient to get the line.

The lines cross at (100, 30) ❸. 100 units of gas are used and the cost is £30.

This means that G-gas is cheaper up to 100 units. At 100 units both companies charge the same amount. After 100 units P-gas is cheaper.

Look out for

When you set up an equation you need to make sure that you use the same units throughout the equations.

Where the lines cross is the solution to the simultaneous equations.

Key terms

Simultaneous equations

Gradient and intercept of straight line graphs

Intersection of two lines

b Here is the graph of the circle $x^2 + y^2 = 9$.

Find the solution to the following simultaneous equations.

$x^2 + y^2 = 9$ and $y + 2x = 2$.

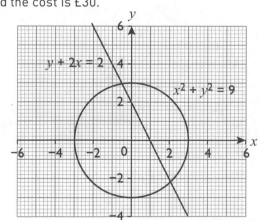

Answer

The solution is where the two lines cross so you need to draw in the straight line $y = -2x + 2$ on the grid.

Intersect at $x = -0.5$, $y = 3.0$ and $x = 2.1$ $y = -2.2$ (correct to 1 d.p.) ❸

Exam-style questions

1 Use a graphical method to find the point where the lines $x + 3y = 2$ and $y = 3x + 4$ cross. **[4]**

2 Here are the tariffs for two mobile phone companies. **M-mobile** charges £20 a month and data costs 50p per Mbyte. **Peach** charges £10 a month and data costs 75p per Mbyte. Explain which company is cheaper. **[5]**

Exam tip

Be careful when you read off the results of the point of intersection of the lines when the scales are different on the two axes.

CHECKED ANSWERS ☐

Solving linear inequalities

Rules

1. Change the inequality into an equation of the form $y = mx + c$.
2. Draw the straight line.
3. Shade according to the inequality either the wanted or unwanted region.
4. Check the shading by substituting the co-ordinates of a point on the grid to check it fits the inequality.
5. Shade the feasible region or shade the unwanted region and label accordingly.

Worked example

a Write down the inequalities that define the feasible region shown on the grid.

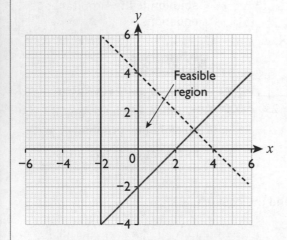

Key terms

Inequality

Less than <

Less than or equal to ⩽

More than >

More than or equal to ⩾

Feasible region

Answers

The vertical line is $x = -2$.

So the inequality is $x \geqslant -2$.

The red slanted line is $y = x - 2$.

So the inequality is $y \geqslant x - 2$.

The dashed slanted line is $x + y = 4$.

So the inequality is $x + y < 4$.

4 You can check the feasible region by substituting in a point in the region, for example, (0, 0) so 0 is ⩾ −2 works as does 0 ⩾ 0 − 2 and 0 + 0 < 4. 2, 3 & 5

b Derek works less than 60 hours. He makes sports boats (s) and rowing boats (r).

It takes 12 hours to make a sports boat and 5 hours to make a rowing boat. He makes more sports boats than rowing boats.

Answers

Show this on the grid:

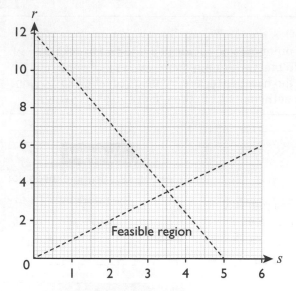

For the hours worked:

❶ $12s + 5r < 60$

For the number of boats:

❶ $s > r$

❷ Draw the lines.

❺ Show the feasible region.

Exam-style question

1 On a co-ordinate grid with x values from −3 to +3 and y from −2 to +3, show the region defined by these inequalities:

$y < x + 1$

$x \leq 1$

$y + 2x > 0$ **[4]**

CHECKED ANSWERS

Factorising quadratics of the form $x^2 + bx + c$

Rules

1. A quadratic expression of the form $x^2 + bx + c$ can sometimes be factorised into two brackets.
2. When the coefficient of x^2 is 1 then each bracket will start with x e.g. $(x \quad)(x \quad)$
3. The number terms in the brackets multiply to give the number term c in the quadratic expression.
4. The number terms in the brackets add to give the coefficient of x in the quadratic expression.

Worked examples

a Factorise $x^2 + 5x + 6$

Answer

The number term is + 6 so the numbers in the brackets multiply to give + 6 which means they could be +1 and +6 **or** –1 and –6 **or** +2 and +3 **or** –2 and –3.

The coefficient of x is 5 so the two numbers need to add up to 5 which means they will need to be 2 and 3.

So $x^2 + 5x + 6 = (x + 2)(x + 3)$ because $x^2 + (2 + 3)x + 2 \times 3$.

b Factorise $x^2 - 5x + 6$

Answer

The number term is + 6 so the numbers in the brackets multiply to + 6 which means they could be +1 and +6 **or** –1 and –6 **or** +2 and +3 **or** –2 and –3.

The coefficient of x is –5 so the two numbers need to add up to –5 which means they will need to be –2 and –3.

So $x^2 - 5x + 6 = (x - 2)(x - 3)$ because $x^2 + (-2 + -3)x + -2 \times -3$.

c Factorise $x^2 - x - 6$

Answer

The number term is –6 so the numbers in the brackets multiply to –6 which means they could be +1 and –6 **or** –1 and +6 **or** +2 and –3 **or** –2 and +3.

The coefficient of x is –1 so the two numbers need to add up to –1 which means they will need to be +2 and –3.

So $x^2 - x - 6 = (x + 2)(x - 3)$ because $x^2 + (+2 + -3)x + +2 \times -3$.

d Factorise $x^2 + x - 6$

Answer

The number term is –6 so the numbers in the brackets multiply to –6 which means they could be +1 and –6 **or** –1 and + 6 **or** +2 and –3 **or** –2 and +3.

The coefficient of x is +1 so the two numbers need to add up to +1 which means they will need to be –2 and +3.

So $x^2 + x - 6 = (x - 2)(x + 3)$ because $x^2 + (-2 + 3)x + -2 \times +3$.

e Factorise $x^2 - 25$. This is called the **difference of two squares**.

Answer

You get $x^2 - 25 = (x + 5)(x - 5)$ because $x^2 + (+5 + -5)x + +5 \times -5$.

Key terms

Quadratic expression

Factorise

Coefficient

Brackets

Look out for

When you have to factorise the difference of two squares $x^2 - y^2$ you get $(x + y)$ and $(x - y)$ so $x^2 - y^2 = (x + y)(x - y)$.

Exam-style questions

1 Factorise $x^2 + 6x + 8$ **[2]**
2 Factorise $x^2 - 2x - 8$ **[2]**
3 Factorise $x^2 + 2x - 8$ **[2]**
4 Factorise $x^2 - 6x + 8$ **[2]**
5 Factorise $x^2 - 16$ **[2]**

CHECKED ANSWERS

Exam tip

After you have factorised a quadratic expression, always multiply out the brackets to make sure you get back to the original expression.

Solve equations by factorising

Rules

1. A quadratic equation has a term in x^2 and to solve it, it must always equal 0
2. Factorise the quadratic function that equals 0
3. Make two equations from the factorised expressions; both equal to 0
4. Solve the two equations to get your solutions.

Worked examples

a Solve $x^2 - 5x = 0$

Answer

Quadratic equation = 0 so you can factorise it into $x(x - 5) = 0$ **1** **2**

So you need to make two equations, either $x = 0$ or $x - 5 = 0$ **3**

This means that either $x = 0$ or $x = 5$ **4**

b Solve $x^2 + 3x - 10 = 0$

Answer

Quadratic equation = 0 so you can factorise it into $(x - 2)(x + 5) = 0$ **1** **2**

So you need to make two equations, either $x - 2 = 0$ or $x + 5 = 0$ **3**

This means that either $x = 2$ or $x = -5$ **4**

c Solve $x^2 - 5x = 14$

Answer

This quadratic equation does not equal 0 so you need to re-arrange it into $x^2 - 5x - 14 = 0$ **1**, which can now be factorised into $(x + 2)(x - 7) = 0$ **2**

Now you need to make two equations either $x + 2 = 0$ or $x - 7 = 0$ **3**

This means that either $x = -2$ or $x = 7$ **4**

d A rectangle's length is 5 cm longer than its width. The area of the rectangle is 24 cm². Find the length and width of the rectangle.

Answer

Let the width be x cm. The length will be $(x + 5)$ cm.

Since the area is 24 cm², the equation $x(x + 5) = 24$ can be formed. This needs to be rearranged by multiplying out the bracket into $x^2 + 5x = 24$. Now you need to make the equation = 0 so that $x^2 + 5x - 24 = 0$

Now you factorise to get $(x - 3)(x + 8) = 0$

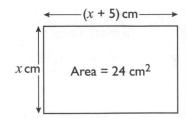

This means $x = 3$ cm or $x = -8$ cm.

As you cannot have a negative length for a measurement, $x = 3$ cm and the answers are length is 8 cm; width is 3 cm.

Look out for

Make sure that the quadratic equation always equals zero. If it does not, then rearrange the equation.

When you have an equation in x^2 then there will be two solutions.

Key terms

Quadratic equation

Solve

Factorise

Exam-style questions

1. Solve
   a $x^2 + 4x - 12 = 0$ **[3]**
   b $x^2 - 5x = 0$ **[3]**
   c $x^2 - 7x = 18$ **[3]**
   d $x^2 - 25 = 0$ **[3]**

2. Ben thinks of a number. He adds 5 to the number and squares his answer. His final answer is 49. What numbers could Ben have been thinking about? **[4]**

Exam tips

You might have to form the equation in order to solve a problem.

Make sure your values for x make sense.

CHECKED ANSWERS

Factorising harder quadratics and simplifying algebraic fractions

Rules

① When the coefficient of the x^2 term is greater than 1 then you will need to check more combinations of factors.

② Use a strategic method to cut down the number of combinations you need to check by looking at the coefficient of x.

③ When you have to simplify algebraic fractions always remember to multiply every term by the common denominator.

Worked example

a Factorise $6x^2 - 7x - 20$.

Answer

The 6 could be made up from 1×6 or 2×3 or -1×-6 or -2×-3.

The 20 could be made up from $-1 \times +20$, or $+1 \times -20$, or $+4 \times -5$, or $-4 \times +5$, or -2×10, or 10×-2.

This means there are 24 different combinations to check.

You can cut this down to 16 if we ignore the -1×-6 and -2×-3.

Since the coefficient of the x term is small, i.e. -7, you can ignore the $-1 \times +20$, $+1 \times -20$, -2×10 and 10×-2 too, as the effect of the 20 would give big numbers as will the 6 when we use x and $6x$.

This leaves us with only four combinations to check.

① $(2x + 4)(3x - 5) = 6x^2 - 10x + 12x - 20 = 6x^2 + 2x - 20$ ✗

$(2x - 4)(3x + 5) = 6x^2 + 10x - 12x - 20 = 6x^2 - 2x - 20$ ✗

$(3x + 4)(2x - 5) = 6x^2 - 15x + 8x - 20 = 6x^2 - 7x - 20$ ✓

$(3x - 4)(2x + 5) = 6x^2 + 15x - 8x - 20 = 6x^2 + 7x - 20$ ✗

So, $6x^2 - 7x - 20 = (3x + 4)(2x - 5)$

b Solve $\frac{3x}{x-3} - \frac{4x}{2x-1} = 1$

Answer

The first step is to multiply everything by $(x - 3)(2x - 3)$ to get:

③ $\frac{3x(x-3)(2x-1)}{x-3} - \frac{4x(x-3)(2x-1)}{2x-1} = 1(x-3)(2x-1)$ then cancel to get

$3x(2x-1) - 4x(x-3) = (x-3)(2x-1)$ then multiply out the brackets

$6x^2 - 3x - 4x^2 + 12x = 2x^2 - x - 6x + 3$ collect like terms

$2x^2 + 9x = 2x^2 - 7x + 3$ simplify to get

$16x = 3$, leaving us with a simple equation

$x = \frac{3}{16}$.

Key terms

Factorise

Coefficient

Combination

Common denominator

Solve

Look out for

Adopt a strategic approach to selecting the combinations to cut down the number of combinations to check.

Make sure that you multiply every term by the common denominator.

Exam tip

Always look for common factors when you have simplify an algebraic fraction.

Exam-style questions

1 The area of this right-angled triangle is $8\,cm^2$.
Calculate the length of the shortest side.
All lengths are in centimetres. **[5]**

2 Simplify $\frac{x-5}{3x^2-11x-20} + \frac{3}{3x+4}$ **[3]**

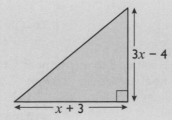

$3x - 4$

$x + 3$

CHECKED ANSWERS

The quadratic equation formula

Rules

1. If a quadratic equation of the form $ax^2 + bx + c = 0$ does not factorise then you can use the formula $x = \frac{-b \pm \sqrt{b^2 - 4ac}}{2a}$ to solve the equation.

2. You substitute the coefficients of the quadratic function $ax^2 + bx + c$ into the formula to get the two solutions.

3. If the discriminant $b^2 - 4ac$ is greater than 0 you will get two answers.

4. If the discriminant $b^2 - 4ac$ equals 0 you will get two equal answers.

5. If the discriminant $b^2 - 4ac$ is less than 0 you will get no answers.

Worked example

a Solve $\frac{2x}{2x-1} - \frac{3x}{4x+1} = 1$. Give your answer correct to 3 decimal places.

Answer

Firstly, you need to clear the fractions by multiplying throughout by the common denominator $(2x - 1)(4x + 1)$. Then cancel to give:

$2x(4x + 1) - 3x(2x - 1) = 1(2x - 1)(4x + 1)$ then multiply out brackets

$8x^2 + 2x - (6x^2 - 3x) = 8x^2 + 2x - 4x - 1$

$8x^2 + 2x - 6x^2 + 3x = 8x^2 + 2x - 4x - 1$ now simplify and collect like terms

$0 = 6x^2 - 7x - 1$ or $6x^2 - 7x - 1 = 0$ then make equation = 0

1. Compare with $ax^2 + bx + c = 0$

So $a = 6$, $b = -7$ and $c = -1$.

2. Substituting these values into the quadratic equation formula gives:

2. $x = \frac{+7 \pm \sqrt{(-7)^2 - 4 \times 6 \times -1}}{2 \times 6} = \frac{7 \pm \sqrt{49 + 24}}{12}$.

The two roots are therefore $x = \frac{7 + \sqrt{73}}{12}$ or $\frac{7 - \sqrt{73}}{12}$.

So $x = 1.295$ or $x = -0.129$.

b Explain why there are no solutions to the quadratic equation $3x^2 - 6x + 10 = 0$.

Answer

5. The value of $b^2 - 4ac$ is $(-6)^2 - 4 \times 3 \times 10 = 36 - 120 = -84$

Since this is less than 0 there are no real solutions.

Key terms

Quadratic equation formula

Coefficients

Substitute

Discriminant

Look out for

If the value of $b^2 - 4ac$ is:

> 0 you get 2 roots

= 0 you get 1 root (equal roots)

< 0 you get 0 roots

You can use the quadratic equation formula if you cannot factorise the quadratic expression.

Exam-style questions

1 Solve $5x^2 - 7x = 3$.

Give your answer correct to 2 decimal places. **[3]**

2 Without solving the equation, explain the number of solutions to the equation $2x^2 - 8x + 20 = 12$. **[3]**

Exam tip

If a question that involves solving a quadratic equation asks you to give the answer correct to a number of significant figures or decimal places, then it is a hint that you need to use the formula.

Using chords and tangents

Rules

1. To find the average rate of change of a variable, you work out the gradient of the straight line joining two points on the graph. For example, to find the average speed over a period of time you draw a right-angled triangle with the hypotenuse as the straight line joining the two points on the graph and divide the vertical distance by the horizontal distance to find the gradient.

2. To find the rate of change at a point, you draw a tangent to the curve at the point and find the gradient. For example, to find the speed when the time is 5 seconds you draw the tangent to the curve at that point and then draw a right-angled triangle with the hypotenuse as the straight line joining the two points on the tangent and divide the vertical distance by the horizontal distance to find the gradient.

Worked example

a The graph shows the speed of a car as it approaches a hazard.

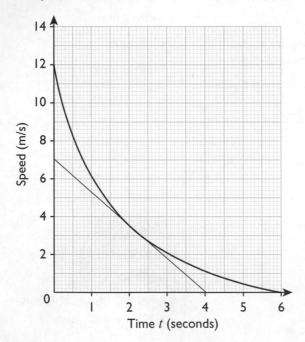

Key terms

Chord

Tangent

Gradient

Rate of change

i Find the average acceleration between $t = 0$ and $t = 5$ seconds.

ii Find the acceleration when $t = 2$ seconds.

Answers

1 i Average acceleration is the chord joining (0, 12) to (5, 0.4)

Gradient is $\dfrac{\text{Difference in } y \text{ readings}}{\text{Difference in } x \text{ readings}} = \dfrac{12 - 0.4}{0 - 5} = \dfrac{11.6}{-5} = -2.32\,\text{ms}^{-2}$

The negative sign means the car is slowing down.

2 ii You find the acceleration at a point drawing a tangent to the curve when $t = 2$.

You select two points on the tangent where it is easy to read off the points.

The points could be (0, 7) and (4, 0).

The gradient is the difference in the ys divided by the difference in the xs.

$= \dfrac{7 - 0}{0 - 4} = \dfrac{7}{-4} = -1.75\,\text{ms}^{-2}$. The negative sign means the car is slowing down.

Look out for

The steeper the gradient then the larger the rate of change.

The shallower the gradient then the smaller the rate of change.

The gradient of a distance–time graph gives velocity.

The gradient of a velocity–time graph gives acceleration.

Exam-style questions

1 The distance travelled by a car when it starts from rest is given by the formula $s = \frac{1}{2}t^2$ where s metres is the distance travelled in time t seconds.
   a Find the speed of the car when $t = 5$ seconds.
   b Find the average speed over the first 5 seconds. **[4]**

Exam tip

Make sure that you read the scales correctly when you find the vertical and horizontal readings as the scales will usually be different.

CHECKED ANSWERS

Translations and reflections of functions

Rules

When the curve of f(x) is:
1. Reflected in the x-axis the function becomes –f(x).
2. Reflected in the y-axis the function becomes f(–x).
3. Translated by the vector $\binom{a}{b}$ the function becomes f(x – a) + b.

Worked example

a Here is the graph of f(x).

Sketch the graph of:

  i   f(–x)
  ii f(x + 2) – 3

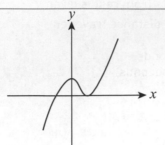

Key terms

Function

Graph

Reflection

Translation

Answers

i

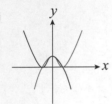

Since it is f(–x) the function is reflected in the y-axis.

ii

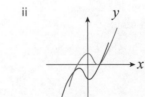

Since it is f(x + 2) – 3 the function moves 2 to the left and 3 down.

b Here is the graph of f(x) = x^2 – 4.

Write down the equations of the lines labelled **P** and **Q**.

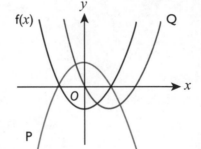

Answers

P is a reflection in the x-axis so its equation is –f(x) or $y = -(x^2 - 4)$ or $y = 4 - x^2$.

Q has been translated 2 to the right as f(x) meets the x-axis at $x = ±2$ so its equation is f(x – 2) + 0 or $y = (x - 2)^2 - 4$ or $y = x^2 - 4x$.

Look out for

If the function has been moved up then b will be positive; if it has been moved down then b is negative.

If the function has moved to the left then a will be positive; if it has been moved to the right then a will be negative.

Exam-style questions

1 f(x) = x^2 – 3x
   Sketch the graphs of:
   i   f(x) **[2]**
   ii f(x + 3) + 2 **[2]**
   iii –f(x) **[2]**

2 The function f(x) = x^3 – 4x is reflected in the y-axis and
   then translated $\binom{2}{-3}$.

   Work out the equation of the resulting function. **[4]**

Exam tip

Make sure that if you have to make a sketch of a transformation of a graph, you write in the key points of the graph.

CHECKED ANSWERS

Area under non-linear graphs

Rules

1. Split the horizontal axis into equal strips. This will form some trapeziums which will all have the same width.
2. Use the area of a trapezium formula to find the area of each trapezium.
3. Add up all the areas to find the total area under the curve.

Worked example

a This graph shows the velocities of a stone and a ball thrown into the air.

Key terms

Area under curve

Speed / velocity

Acceleration

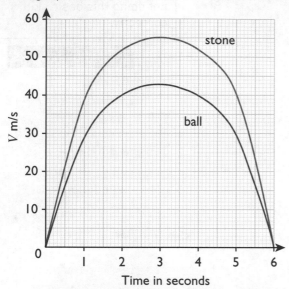

i Find the difference in the distance travelled by the ball and the stone in the 6 seconds shown.

ii Explain why the stone might have travelled faster than the ball.

iii What assumptions have you made?

Answers

i Split the horizontal time axis into six 1 second strips. You have 2 triangles and 4 trapeziums for each graph. **1**

The distance travelled by the stone is:

2 $\frac{1}{2} \times 40 \times 1 + \frac{1}{2} \times (40 + 52) \times 1 + \frac{1}{2} \times (52 + 55) \times 1 + \frac{1}{2} \times (55 + 52) \times 1 + \frac{1}{2} \times (52 + 40) + \frac{1}{2} \times 40 \times 1$

3 $= 20 + 46 + 53.5 + 53.5 + 46 + 20 = 239$ metres

The distance travelled by the ball is:

2 $\frac{1}{2} \times 30 \times 1 + \frac{1}{2} \times (30 + 40) \times 1 + \frac{1}{2} \times (40 + 43) \times 1 + \frac{1}{2} \times (43 + 40) \times 1 + \frac{1}{2} \times (40 + 30) + \frac{1}{2} \times 30 \times 1$

3 $= 15 + 35 + 41.5 + 41.5 + 35 + 15 = 183$ metres

The difference in the distance travelled is $239 - 183 = 56$ metres.

ii The ball has greater air resistance than the stone so will slow down faster.

iii That the ball and the stone had the same initial speed and were launched from the same place.

Look out for

Make sure you remember the area of a triangle.

The area under a speed–time graph gives distance.

The area under an acceleration–time graph gives change in velocity.

Exam-style questions

1 The speed of a runner during the first 5 seconds of a race is given by the formula $v = 2t - \frac{t^2}{5}$.

Plot the graph and find the distance travelled in the first 5 seconds. **[5]**

Exam tip

Make sure that you read the scales off the graphs correctly. Many candidates lose a lot of marks by not doing this basic skill correctly.

CHECKED ANSWERS

Mixed exam-style questions

1 The rectangle has an area of $3x^2 - 12x - 5$.

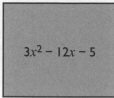

$3x^2 - 12x - 5$

Find the value of x when the area is $15\,cm^2$. Give your answer correct to 3 significant figures. **[3]**

2 The equation $x^3 - 3x - 5 = 0$ has a solution between 2 and 3. Find this solution correct to 3 decimal places. **[2]**

3 a Show that $\dfrac{5}{n-2} - \dfrac{2}{n+2} \equiv \dfrac{3n+14}{n^2-4}$ **[3]**

   b For which values of n is this not true? **[2]**

4 Here is a right-angled triangle. It has an area of $25.5\,cm^2$. The length of the base is $(x + 5)\,cm$. The height is $(x - 3)\,cm$.

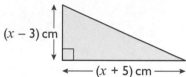

Work out the perimeter of the triangle. Give your answer to 3 significant figures. **[6]**

5 Here is a speed–time graph for a car travelling between two traffic lights.

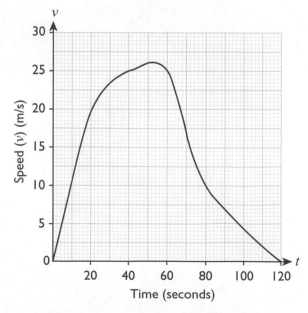

   a Work out the average acceleration in the first 40 seconds. **[1]**
   b Estimate the acceleration when $t = 80$ seconds. **[1]**
   c Estimate the distance between the two sets of traffic lights. **[1]**

6 Find the co-ordinates where the line $y = 2x - 8$ intersects with the circle $x^2 + y^2 = 29$. **[2]**

7 Here are the population figures of stick insects in a vivarium.

Time (t weeks)	0	1	2	3	4	5	6
Population (P)	5	7	10	14	20	28	40

  a Explain why the population increase is exponential. **[2]**
  b Find the equation of the graph. **[3]**

Geometry and Measures: pre-revision check

Check how well you know each topic by answering these questions. If you get a question wrong, go to the page number in brackets to revise that topic.

1 The density of platinum is 21.4 g/cm³.
A wedding ring has a volume of 0.7 cm³.
Calculate the mass of the ring.   (Page 59)

2 Which two of the following triangles are congruent?
Give a reason for your answer.   (Page 60)

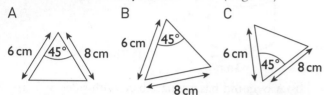

3 Explain why all equilateral triangles are always similar but not all equilateral triangles are congruent.   (Page 61)

4 For each of the diagrams below, find the size of the angle marked with a letter, give reasons for your answers.   (Page 62)

a   **b**

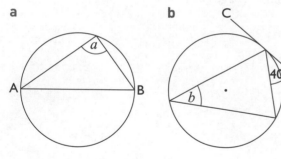

AB is the diameter of the circle. CD is a tangent to the circle.

5 Find the length of the side marked x in this right-angled triangle.   (Page 63)

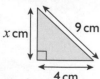

6 AOB is a sector, radius 7 cm. The angle AOB is 120°.
a Calculate the arc length of the sector AOB.
b Calculate the area of the sector AOB.   (Page 64)

7 ABC is a triangle. AB = 7 cm, BC = 10 cm and angle ABC is 112°.
Calculate the length of AC.   (Page 65)

8 Work out the length of the side AB.   (Page 66)

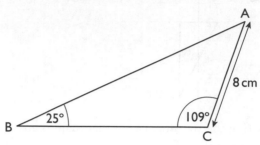

9 a Draw the locus of all the points 3 cm from a point A.
b Draw the locus of all the points equidistant from a pair of parallel lines 4 cm apart.
(Page 68)

10 Triangles A and B are similar.

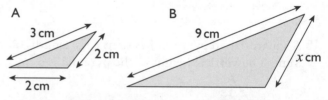

Calculate the value of x.   (Page 71)

11 Work out the length of AC.   (Page 72)

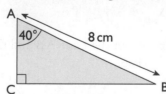

12 Calculate the exact length of AC in the triangle ABC.   (Page 72)

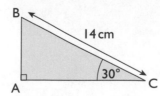

13 Describe fully the transformation that maps triangle T onto triangle R in the diagram below. (Page 73)

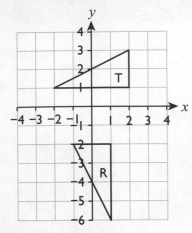

14 Describe the transformation that maps shape A onto shape B. (Page 75)

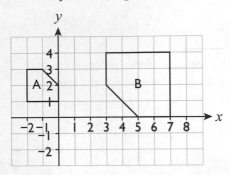

15 A square-based pyramid has slant edge lengths of 8 cm. The vertical height of the pyramid is 6 cm.

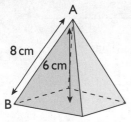

Calculate the angle between the edge AB and the base of the pyramid. (Page 77)

16 Draw the plan, front and side elevations of the object shown below. (Page 81)

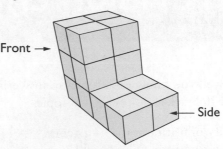

Front →

Side

17 a Calculate the volume and surface area of this cylinder.

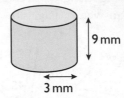

9 mm

3 mm

b A pyramid has a square base with sides of 7 cm. The height of the pyramid is 10 cm. Calculate the volume of the pyramid. (Page 82)

18 Two cuboids are similar. The larger cuboid has a volume of 192 cm³ and a surface area of 208 cm². The smaller cuboid has a volume of 24 cm³. Work out the surface area of the smaller cuboid. (Page 83)

Working with compound units and dimensions of formulae

REVISED

LOW

Rules

❶ A compound measure involves two quantities, for example speed = distance ÷ time.
❷ In a compound unit 'per' means 'for each' or 'for every'.
❸ If you need to change the units of a compound unit change one quantity at a time.
❹ Density is a compound unit. Density = mass ÷ volume.

Worked examples

a The volume of a silver bar is 50 g. The density of silver is 10.5 g/cm³. Work out the mass of the silver bar.

Answer

Density = mass ÷ volume ❹
Density × volume = mass (make mass the subject of the formula) ❶
Mass of bar is $10.5 \times 50 = 525$ g

b A cheetah can run at 33 m/s

  i How long would the cheetah take to run a km?
  ii Work out the cheetah's speed in km/h.

Answer

  i Time = distance ÷ speed ❷
     $= 1000 \div 33 = 30.3$ s
  ii Speed = 0.033 m/s (change m to km divide by 1000) ❸
     $0.033 \times 3600 = 119$ km/h 3 s.f. (multiply by 3600 to change from seconds to hours)

c p, q and r represent lengths. State whether each of the following represents a length, an area or a volume.
  i $3p + \pi r$ **ii** $p(r + q)$ **iii** $r(3p + q)^2$

Answer

  i A length as it is a length added to a length.
  ii $(r + q)$ is two lengths added, which is still a length. When a length $(r + q)$ is multiplied by a length it is an area.
  iii $(3p + q)$ is a length. Squaring means it is an area (length × length). When multiplied by r it becomes a volume.

Key terms

Rate of change

Speed

Density

Population density

Mass

Unit price

Look out for

Remember to change the units.

Exam-style questions

1 The area of Costa Rica is 51 100 km².
In 2013 the population of Costa Rica was 4.87 million. In 2015 the population of Costa Rica was 5.06 million.

Calculate the change in population density of Costa Rica between 2013 and 2015. **[2]**

2 An iron nail has a volume of 0.9 cm³ and a mass of 7 g.
  a Calculate the density of iron. **[1]**
  b An iron girder has a volume of 1.5 m³.
    Calculate the mass of the girder in kg. **[1]**

3 A motorway traffic sign indicates that it is 13 mins to the next junction on the motorway.
The junction is 14 miles away.

What assumption has been made about the speed of the traffic? **[3]**

Exam tips

Population density is a compound measure so the units for your answer should include two quantities.

Make sure you show all the stages of your working.

CHECKED ANSWERS

Congruent triangles and proof

Rules

Two triangles are congruent if one of the following conditions is true:
1. The three sides of each triangle are equal (SSS).
2. Two sides and the included angle are equal (SAS).
3. Two angles and the corresponding side are equal (AAS).
4. Each triangle contains a right angle, and the hypotenuse and another side are equal (RHS).

Worked examples

a Which of the following triangles are congruent? Give reasons for your answers.

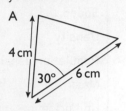
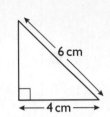
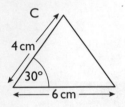
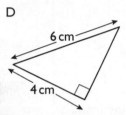

Key terms

Congruent

Proof

Answer

A and C are similar; two sides and the included angle are equal. ②

B and D are congruent, they are right-angled triangles and the hypotenuses and one other side are equal. ④

Exam tip

Make sure you know all the conditions for congruency.

b PQRS is a rhombus. Prove that triangles PQX and RSX are congruent.

Answer

PQ = RS, all sides of a rhombus are equal.

PX = XR and SX = XQ, diagonals of a rhombus bisect each other. ①

So triangles PQX and RSX are congruent (SSS).

Exam tip

Make sure you finish your proof with a conclusion.

Exam-style questions

1 ABC is an equilateral triangle. X and Y are the midpoints of sides AB and BC. Prove that AYC and AXC are congruent triangles. **[3]**

2 PQRS is a parallelogram. X is the midpoint of PQ. Y is the midpoint of RS. Prove XQS and QYS are congruent. **[3]**

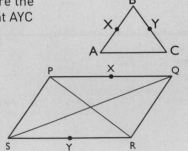

Exam tip

Only use the properties given in the question in your answer

Proof using similar and congruent triangles

Rules

1 Shapes are similar if one is an enlargement of the other.

Two triangles are congruent if one of the following conditions is true:
2 The three sides of each triangle are equal (SSS).
3 Two sides and the included angle are equal (SAS).
4 Two angles and the corresponding side are equal (AAS).
5 Each triangle contains a right angle, and the hypotenuse and another side are equal (RHS).

Worked examples

a Which of the following triangles are similar? Give reasons for your answers.

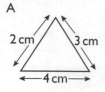 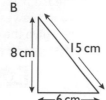 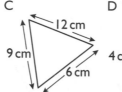 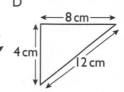

Key terms

Similar

Congruent

SSS, SAS, ASA, RHS

Answer

A and C are similar, the lengths of the sides of triangle C are all three times the lengths of the sides of triangle A. **1**

b Prove that triangles WXY and WZY are congruent.

Answer

WY is the hypotenuse of both triangles. WX = ZY (given); angles WZY and WXY are right angles (given); so WXY and WZY are congruent (RHS). **5**

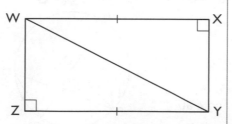

Exam tip

Many students have problems with these types of questions because they have not learnt the rules for congruent triangles.

Exam-style questions

1 Prove that triangles AXD and BXC are similar. **[3]**

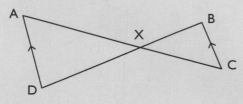

Exam tip

The only properties you can use to prove congruency or similarity are those given in the question.

Make sure you show every step of your reasoning.

2 PQRST is a regular pentagon. Prove that triangles QRS and STP are congruent, hence prove that PQS is an isosceles triangle. **[5]**

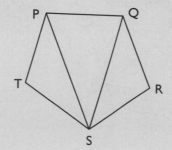

CHECKED ANSWERS

Circle theorems

Rules

❶ The angle at the centre of a circle is twice the angle at the circumference on the same arc.
❷ The angle in a semicircle is 90°.
❸ Angles in the same segment are equal.
❹ Opposite angles of a cyclic quadrilateral add up to 180°.
❺ The line joining the centre of a circle to the midpoint of a chord is at right angles to it.
❻ The angle between a radius and a tangent is 90°.
❼ The angle between a chord and a tangent is equal to the angle in the alternate segment.

Worked examples

a For each of the diagrams below find the size of the angle marked with a letter, give reasons for your answers.

i 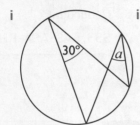 ii 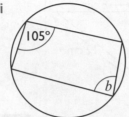 iii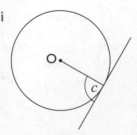

Answers

i $a = 30°$, angle a is in the same segment as the angle 30°. ❸

ii $b = 75°$, opposite angles in a cyclic quadrilateral add up to 180°. ❹

iii $c = 90°$, the angle between a radius and a tangent is 90°. ❻

b In the circle, centre O, BOC = 50°.
Show that BDC = 25°.

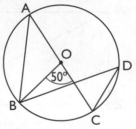

Answer

Angle BAC = 25°. The angle at the centre of a circle is twice the angle at the circumference. ❶

BDC = 25°, angles in the same segment are equal. ❸

Look out for

When you use circle theorems as reasons for your answers, make sure you give the full theorem.

Key terms

Semi circle

Segment

Cyclic quadrilateral

Chord

Tangent

Alternate segment

Exam tip

When you have to use several steps to solve a problem it can help to mark the diagram with information as you find it.

Exam-style questions

1 Find the size of the angles marked with a letter, give reasons for your answers.
Where shown O is the centre of the circle.

2 P, Q and R are points on the circumference of the circle. QT is a tangent to the circle. QR bisects the angle PQT.
Prove that RP = RQ. **[3]**

a **[3]**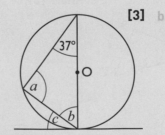

b **[2]**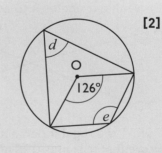

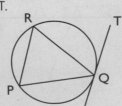

Pythagoras' theorem

Rules

❶ Pythagoras' theorem can be written as $a^2 + b^2 = h^2$, where h is the hypotenuse of a right-angled triangle and a and b are the other sides of the triangle.

❷ To find h add the squares of each of the other sides together and then square root the answer.

❸ To find a or b subtract the square of the known side from the square of the hypotenuse and square root the answer.

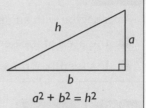

$a^2 + b^2 = h^2$

Worked examples

a ABC is a right-angled triangle.
Find the length of side AB to 1 d.p.

Answer

$AB^2 = AC^2 + CB^2$ (AB is hypotenuse) ❶ ❷

$AB^2 = 9^2 + 7^2$

$AB^2 = 81 + 49$

$AB = \sqrt{130}$ (use the square root button
to find $\sqrt{130}$)

$AB = 11.4$ cm to 1 d.p.

b Find the value of x.

Answer

$x^2 = 12.3^2 - 8.5^2$ (x is a smaller side) ❸

$x^2 = 151.29 - 72.25$

$x = \sqrt{79.04}$

$x = 8.9$ cm to 1 d.p.

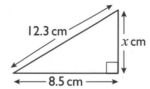

Key terms

Right-angled triangle

Pythagoras

Hypotenuse

Square

Square root

Look out for

Make sure you write down all the steps in your working.

Exam tip

The hypotenuse of a right-angled triangle is the longest side. Always identify the hypotenuse before you start answering the question.

Exam-style questions

1 PQR is an isosceles triangle.

PQ = PR = 8 cm

QR = 5 cm

Calculate the vertical height of the triangle PQR, give your answer to 1 d.p. **[3]**

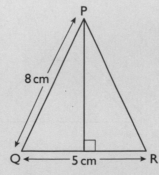

Exam tip

Draw a diagram of the course and mark it with the information given to you in the question.

2 Sally sails a yacht around a course marked by three coloured buoys. The red buoy is 700 metres east of the blue buoy. The green buoy is 1500 metres south of the blue buoy. Calculate the total length of the course. Give your answer to the nearest metre. **[3]**

CHECKED ANSWERS ☐

Arcs and sectors

Rules

For a sector with angle $\theta°$ of a circle and radius r:

❶ the area of the sector is $\frac{\theta}{360} \times \pi r^2$

❷ the length of the arc is $\frac{\theta}{360} \times \pi d$ or $\frac{\theta}{360} \times 2\pi r$

Worked examples

a i Find the area of the shaded sector.

   ii Find the arc length of the shaded sector.

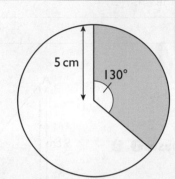

Answer

i area of the sector $= \frac{\theta}{360} \times \pi r^2$ **❶**

$= \frac{130}{360} \times \pi \times 5^2$

$= 28.4 \text{ cm}^2$ to 2 d.p.

ii arc length $= \frac{\theta}{360} \times 2\pi r$ **❷**

$= \frac{130}{360} \times 2 \times \pi \times 5$

$= 11.3 \text{ cm}$ to 2 d.p.

b A sector of a circle with radius 6 cm has an arc length of 8 cm. Find the angle of the centre of the sector to the nearest degree.

Answer

arc length $= \frac{\theta}{360} \times 2\pi r$

$8 = \frac{130}{360} \times 2 \times \pi \times 6$

$2880 = 12\pi x$

$2880 \div 12\pi = x$, so the angle at the centre $= 76°$ to nearest degree.

Look out for

These types of questions are often poorly answered because learners do not learn the formulae in these rules.

Key terms

Radius

Diameter

Circumference

Sector

Arc

Chord

Segment

Exam-style questions

1 The diagram shows a sector of a circle, centre O.

   The radius of the circle is 9 cm.

   Angle POQ is 80°.

   Work out the perimeter of the sector to the nearest cm. **[4]**

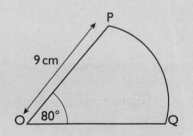

2 The diagram shows the design for a pendant.

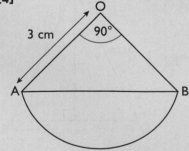

The pendant is made from a gold isosceles triangle joined to a silver segment to form a sector. The finished pendant is a sector AOB radius 3 cm with a centre angle of 90°. Calculate the area of the silver part of the pendant. **[5]**

The cosine rule

Rules

1. The cosine rule is $a^2 = b^2 + c^2 - 2bc \cos A$.
2. It can be used to find the third side of a triangle if you know two sides and the angle between them.
3. It can also be used to find an angle in a triangle if you know the lengths of all the sides. ◄──────

The cosine rule can also be written as:

$b^2 = a^2 + c^2 - 2ac \cos B$

or

$c^2 = a^2 + b^2 - 2ab \cos C$

Worked examples

a The diagram shows the triangle ABC.

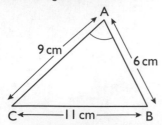

Calculate the size of the angle at A.

Answer

$a^2 = b^2 + c^2 - 2bc \cos A$ ❶

$11^2 = 9^2 + 6^2 - 2 \times 9 \times 6 \cos A$ ❷

$\cos A = \frac{-4}{108}$

$A = 92°$

b The diagram shows an isosceles trapezium.

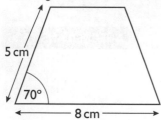

Calculate the length of one of the diagonals of the trapezium.

Answer

$a^2 = b^2 + c^2 - 2bc \cos A$ ❶

$a^2 = 8^2 + 5^2 - 2 \times 8 \times 5 \cos 70°$ ❸

$a^2 = 61.638 \ldots$

$a = 7.85$

So the diagonal of the trapezium is 7.85 cm (2 d.p.)

Look out for

When you use the cosine rule, make sure the angle in the left-hand side of the equation is the angle opposite the side on the right-hand side of the equation.

Exam-style questions

1 The diagram shows a step ladder opened at a hinge to form an isosceles triangle.

3 m

1.8 m

The sides of the ladder are 3 m long. The feet of the ladder are 1.8 m apart.

Calculate the angle at which the hinge has been opened. **[3]**

2 A plane flies a circular route between London, Amsterdam and Paris.

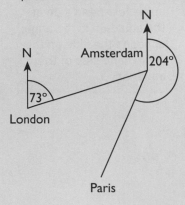

Amsterdam is 358 km from London on a bearing of 073°.

Paris is 510 km from Amsterdam on a bearing of 204°.

Calculate the distance between Paris and London. **[5]**

> **Exam tip**
>
> It can be helpful to label the sides and angles in a diagram before you start answering the question.

CHECKED ANSWERS

The sine rule

HIGH

Rules

① The sine rule can be written as $\frac{\sin A}{a} = \frac{\sin B}{b} = \frac{\sin C}{c}$ or $\frac{a}{\sin A} = \frac{b}{\sin B} = \frac{c}{\sin C}$.

② The sine rule can be used for finding a side when you know 2 angles and another side.

③ The sine rule can be used to find an angle when you know 2 sides and another angle.

④ The area of a triangle can be found using the formula $\frac{1}{2}ab \sin C$.

Worked examples

a ABC is a triangle. AB = 11 cm, AC = 15 cm.
Angle ABC = 65°.

   i Calculate the size of angle ACB.

   ii Calculate the length of side BC.

Answer

i $\frac{\sin C}{c} = \frac{\sin B}{b}$ **①**

$\frac{\sin C}{11} = \frac{\sin 65°}{15}$ **②**

$\sin C = \frac{\sin 65° \times 11}{15}$

BCA = 41.7°, so A = 73.3°

ii $\frac{a}{\sin A} = \frac{b}{\sin B}$

$\frac{a}{\sin 73.3°} = \frac{15}{\sin 65°}$ **③**

$a = \frac{15 \times \sin 73.3°}{\sin 65°}$

BC = 15.9 cm

b Calculate the area of the triangle PQS.

Answer

Area of triangle $= \frac{1}{2}ab \sin C.$ **④**

$= \frac{1}{2} \times 6 \times 9 \times \sin 43°$

$= 18.4 \text{ cm}^2$

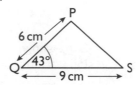

Key facts

Use $\frac{\sin C}{c} = \frac{\sin B}{b}$ to find an angle.

Use $\frac{a}{\sin A} = \frac{b}{\sin B}$ to find a side.

Look out for

Make sure you use opposite sides and angles with the sine rule.

Exam tip

Use the full values from your calculator, do not round too early.

Exam-style questions

1 Zeta is planning a walk between 3 towns, Weston, Easthorpe and Southam.

  Easthorpe is on a bearing of 070° from Weston. Southam is 9 km from Easthorpe on a bearing of 117°. The distance between Weston and Southam is 15 km.

  Calculate the bearing of Southam from Weston. **[4]**

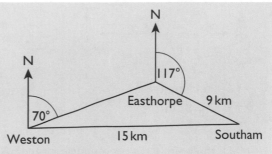

2 A regular hexagon has sides of length 6 cm.

  Calculate the exact area of the hexagon. **[3]**

3 ABCDE is a regular pentagon, with sides 5 cm.

  Point O is the centre of the pentagon.

  a Calculate the length OA. **[4]**

  b Calculate the area of the pentagon ABCDE. **[3]**

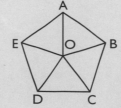

Loci

LOW

Rules

❶ A locus is a line or a curve that joins all the points that obey a given rule.

Some of the most common loci are:
❷ A constant distance from a fixed point
❸ Equidistant from two given points
❹ A constant distance from a given line
❺ Equidistant from two lines

Worked examples

a Draw the locus of all the points that are equidistant from two points, A and B, which are 5 cm apart.

Answer

Ax
_____
Bx

The locus is the perpendicular bisector of the line joining A and B. ❶ ❸

b A horse is tethered by a rope to a straight rail 10 m long. The rope is 1.5 m long. The horse can walk around both sides of the rail. Make a sketch of locus of the maximum reach of the horse. ❶ ❹

Answer

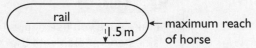
rail
1.5 m ← maximum reach of horse

Key terms

Locus

Loci

Equidistant

Exam-style questions

1 The position of two mobile phone masts A and B are shown. The signal from mast A can reach 10 km. The signal from mast B can reach 7 km.
Using a scale of 1 cm to 2 km, draw an accurate diagram to show the region covered by both masts. **[3]**

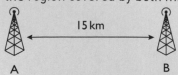

15 km

A   B

2 XYZ is a triangle.

XY = YZ = 8 cm

XZ = 6 cm

Mark the point x that is equidistant from X, Y and Z. **[4]**

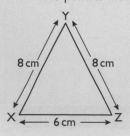

8 cm   8 cm

X ◄— 6 cm —► Z

Exam tips

If you are asked to draw an accurate diagram, you should use a ruler and/or compass to construct the loci.

You should leave all the construction marks.

Remember

Do not forget to make it clear to the examiner where the final region required is by shading and labelling it.

CHECKED ANSWERS

Mixed exam-style questions

1 ABC is an isosceles triangle. D and E are points on BC. AB = AC, BD = EC.

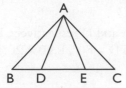

Prove that triangle ADE is isosceles.   [3]

2 In the diagram, PQ is parallel to ST. QX = XS.

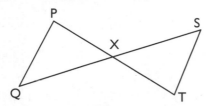

Prove that triangles PQX and STX are congruent.   [3]

3 The diagram shows a right-angled triangle drawn inside a quarter circle. The chord AC is 3 cm.

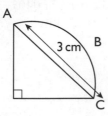

   a Calculate the radius of the circle.   [3]
   b Calculate the area of the segment ABC.   [4]

4 A paper cone is made from a folding a piece of paper in the shape of sector of a circle. The angle at the centre of the sector is 100°. The radius of the sector is 6 cm.

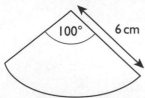

   a Calculate the length of the arc of the sector.   [2]
   b Calculate the diameter of the base of the finished cone.   [2]

5 The diagram shows a trapezium PQRS. PQ is parallel to SR. PS = QR.

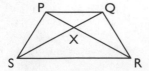

Show that triangles PXQ and SXR are similar.   [3]

6 ABCD is a rhombus. AC = 10.8 cm, BD = 15.6 cm.

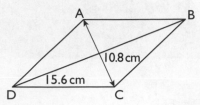

Calculate the length of the sides of the rhombus. **[4]**

7 In the diagram TP and TS are tangents to the circle touching the circumference at A and D respectively. B and C are points on the circumference of the circle. AC and BD cross at point X. ATD = 80°, PAB = 20°.

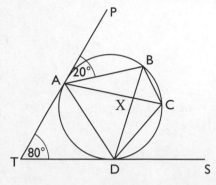

a Show that triangle AXB and triangle DXC are similar triangles. **[3]**
b Prove that ABCD is a cyclic quadrilateral. **[4]**

8 A ferry sails between three islands, A, B and C. Island B is 50 km from island A on a bearing of 040°. Island C is on a bearing of 095° from island B. The distance from island A to island C is 120 km.

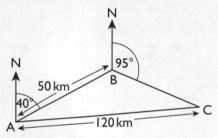

Calculate the bearing of island A from island C. **[5]**

9 The diagram shows a wedge in the shape of a triangular prism.

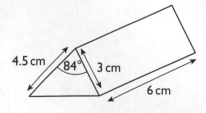

a Calculate the volume of the prism. **[3]**
b Calculate the surface area of the prism. **[4]**

Similarity

Rules

1. Shapes are similar if one shape is an enlargement of the other.
2. The lengths of the sides of similar shapes are in the same ratio.
3. All the angles in similar shapes are the same.

Worked examples

a The diagram shows two similar triangles.

   i Calculate the value of x.

   ii Calculate the value of y.

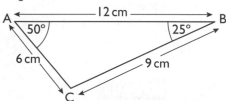

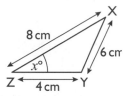

Answer

  i The ratio of the hypotenuses is $2:1$ **2**

    So $x = 2 \times 4.5$

      $x = 9$

  ii $y = 12 \div 2$ **2**

     $y = 6$

b The diagram shows two triangles, ABC and XYZ. Find the size of angle XZY.

Answer

The corresponding sides of triangles ABC and XYZ are in the same ratio. **1** **2**

Therefore triangles ABC and XYZ are similar triangles. Angle XZY corresponds to angle BAC. **3**

So $x = 50$

Key terms

Similar

Ratio

Remember

You can work out the ratio by dividing the lengths of corresponding sides.

Look out for

When answering questions on similar shapes, make sure you use the ratio of corresponding sides.

Exam-style questions

1. The diagram shows triangle ABC.
   Angle BAC = 37°
   AC = 10.5 cm, BC = 7 cm
   YC = 4.5 cm
   The line XY is parallel to the line BC.
   Calculate the length of the line XY. **[3]**

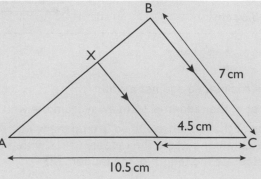

2. PQ is parallel to SR.
   QP = 8 cm
   SR = 10 cm
   PX = 6 cm
   Calculate
   length of PS. **[3]**

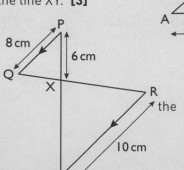

Trigonometry

Rules

For a right-angled triangle:

❶ $\tan\theta = \dfrac{\text{opposite}}{\text{adjacent}}$

❷ $\sin\theta = \dfrac{\text{opposite}}{\text{hypotenuse}}$

❸ $\cos\theta = \dfrac{\text{adjacent}}{\text{hypotenuse}}$

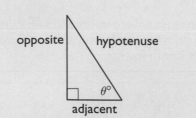

opposite   hypotenuse

$\theta°$

adjacent

Key terms

Sine

Cosine

Tangent

Worked examples

a ABC is a right-angled triangle.

Work out the size of angle ACB to 1 d.p.

Answer

$\cos B = \dfrac{8.6}{13.7}$ ❸

$\cos B = 0.6277$

ACB = 51.1° (Use the inverse cos on your calculator).

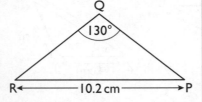

A

8.6 cm

B ← 13.7 cm → C

b PQR is an isosceles triangle.

Angle PQR = 130°, QR = PQ

PR = 10.2 cm

Calculate the length of QR to 1 d.p.

Answer

As PQR is isosceles, RXQ is a right-angled triangle.

RX = 5.1 cm, RQX = 65°

$\sin 65° = \dfrac{5.1}{QR}$ ❷

$QR = \dfrac{5.1}{\sin 65°}$

QR = 5.63 cm

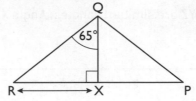

Q

130°

R ← 10.2 cm → P

Q

65°

R ← → X   P

Look out for

Although you use a calculator for these problems, you still need to show the stages of your working.

Exam tip

It is often helpful to mark new information on a diagram, or even draw a new one.

Exam-style questions

1 A ladder is placed against a wall.

The foot of the ladder is 1.2 m away from the wall.

The top of the ladder is placed 3 m up the wall.
  a Calculate the angle the ladder makes with the wall. **[2]**
  b Calculate the length of the ladder. **[2]**

2 PQRS is a rectangle.

PQ = 8 cm and angle QPR = 50°.

Use trigonometry to calculate the length of the diagonal, PR. **[3]**

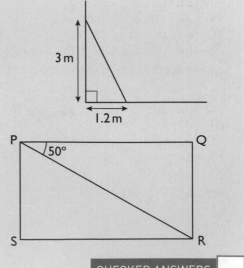

3 m

1.2 m

P   50°   Q

S   R

Finding centres of rotation

Rules

1. To describe a rotation fully you need to state the direction, angle and centre of rotation.
2. The centre of rotation is where the perpendicular bisectors of the lines that join the corresponding points of the image and object cross.

Worked examples

a Describe the rotation that maps

  i   A → B
  ii A → C
  iii A → D

Answer

  i   180° clockwise (or anti-clockwise) rotation about (0,0) **1**

  ii 90° clockwise rotation about (0,0) **1**

  iii 90° anti-clockwise rotation about (0,0) **1**

b Image P has been rotated to form image Q.

  i   Find the centre of rotation that maps P to Q.

  ii Describe fully the transformation.

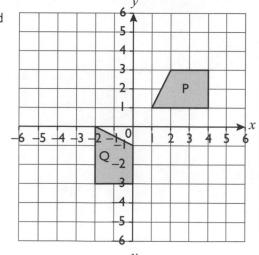

Answer

  i   Centre of rotation is (−1,2) **1**

  ii 90° clockwise rotation about (−1,2) **2**

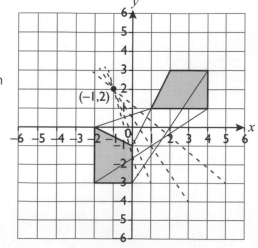

Key terms

Transformation

Rotation

Centre of rotation

Clockwise

Anti-clockwise

Exam tip

You should leave the lines you used to find the centre of rotation in your answer, as you may get some marks for them if you make a mistake later.

Look out for

Don't forget to write down your answer after finding the centre of rotation.

1 Quadrilateral A has been rotated to position B.
   a   Find the centre of rotation that maps quadrilateral A
       to quadrilateral B. **[2]**
   b   Describe fully the transformation that maps
       quadrilateral A to quadrilateral B. **[2]**

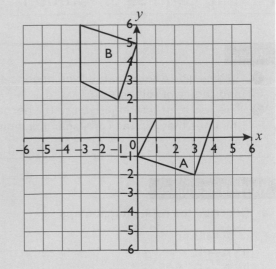

2 The diagram shows two triangles A and B. Describe fully
  the transformation that maps A onto B. **[2]**

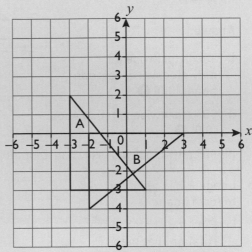

CHECKED ANSWERS

Enlargement with negative scale factors

Rule

❶ When a shape is enlarged by a negative scale factor the image appears on the opposite side of the centre of enlargement and is rotated by 180°.

Key terms

Enlargement

Negative scale factor

Worked examples

a Describe the enlargement that maps:

  i shape A to shape B

  ii shape B to shape A.

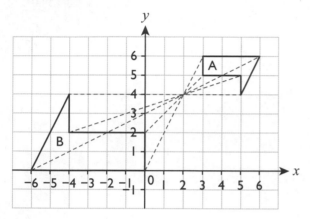

Answers

  i Enlargement with scale factor –2, centre of enlargement (2, 4). ❶

  ii Enlargement with scale factor $-\frac{1}{2}$, centre of enlargement (2, 4). ❶

b Triangle K has co-ordinates (2, 0), (3, 2) and (2, 2).

  i Draw triangle K on a set of axes.

  ii Enlarge triangle K by a scale factor of –3, centre of enlargement (1, 1). Label the new triangle L. ❶

Answers

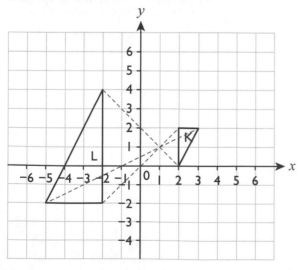

Exam tip

The centre of enlargement is where the lines which join the corresponding points of the image and object cross.

Look out for

To describe an enlargement fully you need to state both the scale factor and centre of enlargement.

Exam-style questions

1 Describe the transformation that maps shape P to shape T. **[2]**

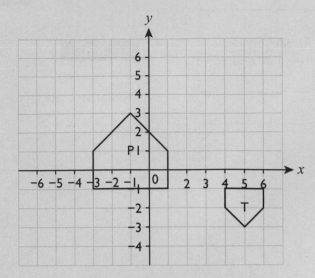

2

a Enlarge rectangle S by a scale factor of $-\frac{1}{3}$, centre of enlargement (1, –1). Label the new rectangle T. **[2]**

b Enlarge rectangle T by a scale factor of –2, centre of enlargement (1, –1). Label the new rectangle U. **[2]**

c Describe the single transformation that maps rectangle S to rectangle U. **[1]**

3 Describe fully the single transformation that has the same effect as an enlargement of scale factor –1, centre (x, y). **[2]**

CHECKED ANSWERS

Trigonometry in 2D and 3D

Rule

❶ To solve problems involving 3D shapes you need to identify relevant 2D triangles.

❷ The angle between a line and a plane is the angle between the line and its projection onto this plane.

Worked examples

a The diagram shows a square-based pyramid ABCDE.

The vertical height, AO, of the pyramid is 10 cm.

BC = CD = DE = EB = 4 cm.

Calculate the angle between the plane ABC and the base of the pyramid.

Exam tip

Always make a drawing of the triangles you are using to solve the problem.

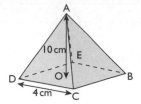

Answer

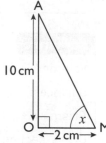

The angle between plane ABC and the base is the angle between AM and MO, where M is the mid-point of BC. ❶ ❷

$$\tan x° = \frac{10}{2}$$

$$x = 78.7°$$

b The diagram shows a cuboid, ABCDEFGH.

Calculate the angle DFH.

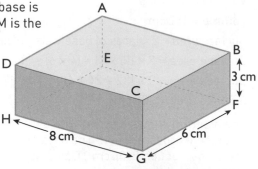

Answer

$$HF^2 = HG^2 + GF^2 \ ❶$$

$$HF^2 = 100$$

$$HF = 10 \text{ cm}$$

$$\tan DFH = \frac{3}{10}$$

$$DFH = 16.7°$$

Look out for

This type of question often requires you to solve the problem in several steps. Make sure you show all the steps and intermediate results.

Exam-style questions

1 ABCD is a regular tetrahedron. All the edges of the tetrahedron are 5 cm long.

Calculate the angle between the planes ABC and BCD. **[7]**

2 The diagram shows a cube with sides of 7 cm.
   a   Calculate the angle the diagonal SU makes with the base of the cube. **[5]**
   b   Calculate the angle between the planes PRX and the base of the cube. **[3]**

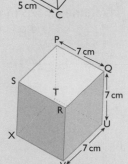

3 The angle between the slant height of a cone and its base is 75°.
The slant height of the cone is 10.8 cm

Calculate the radius of the base of the cone. **[3]**

CHECKED ANSWERS

Volume and surface area of cuboids and prisms

Worked examples

a Work out the volume and surface area of this cuboid.

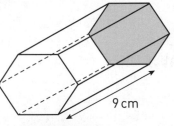

Answer

Volume = length × width × height ❶

Volume = 8 × 4 × 6

Volume = 192 cm³

Surface area = 2 × (area of base + area of side + area of front) ❷

Surface area = 2 × ((8 × 4) + (4 × 6) + (8 × 6))

Surface area = 2 × (32 + 24 + 48)

Surface area = 208 cm²

b The diagram shows a prism with a hexagonal cross-section. The area of the cross-section is 36 cm². Calculate the volume of the prism.

Answer

Volume of a prism = area of cross section × length ❸

Volume = 36 × 9 = 324 cm³

c A cylinder has a radius of 4 mm and length 11 mm. Workout the volume and surface area of the cylinder.

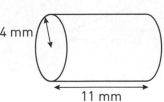

Answer

Volume of a cylinder
= $\pi r^2 h$ ❺

Volume = π × 4 × 4 × 11

Volume = 552.9 mm²
(1 d.p.)

Surface area of a cylinder
= $2\pi rh + 2\pi r^2$ ❻

Surface area = (2 × π × 4 × 11) + (2 × 50.26...)

Surface area = 274.6...+ 100.53

Surface area = 377.0 mm³ (1 d.p)

Exam tip

Don't forget to include units in your answers.

Key terms

Volume

Surface area

Prism

Cross-section

Faces

Remember

The **cross-section** of a prism is the shape you get when you cut the shape at right angles to its length.

Look out for

When you use a calculator to work out problems do not round your answers until the final answer.

Exam-style questions

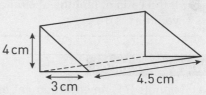

1 A cylindrical water tank has a radius of 2.5 m and a height of 5 m.
   a   Calculate the volume of the tank. **[3]**
   b   Workout the maximum amount of water the tank can hold in litres. **[2]**

2 The diagram shows a triangular prism. The cross-section of the prism is
   a right-angled triangle.
   a   Calculate the area of the cross-section of the prism. **[2]**
   b   Calculate the volume of the prism. **[1]**
   c   Calculate the surface area of the prism. **[3]**

3 A manufacturer makes stock cubes. The stock cubes are made
   in 2 cm cubes. She wants to sell the cubes in boxes of 12 and they
   will be packed with no spaces.
   Work out the dimensions of all the possible boxes the manufacturer could choose from. **[3]**

CHECKED ANSWERS

Enlargement in two and three dimensions

Rules

1. If the ratio of lengths of similar shapes is $1 : x$, the ratio of their areas is $1 : x^2$
2. If the ratio of lengths of similar shapes is $1 : x$, the ratio of their volumes is $1 : x^3$

Worked examples

a Triangle P is an enlargement of triangle Q. The area of triangle Q is $3\,cm^2$. Calculate the area of triangle P.

Answer

Ratio of sides = 2 : 6 or 1 : 3

Ratio of area = $1 : 3^2$ or 1 : 9 ❶

Area of triangle P = $9 \times 3\,cm^2$

$= 27\,cm^2$

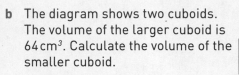

Q

2 cm

P

6 cm

b The diagram shows two cuboids. The volume of the larger cuboid is $64\,cm^3$. Calculate the volume of the smaller cuboid.

Answer

Ratio of sides = 2:4 or 1:2

Ratio of volumes = $1:2^3$ or 1:8 ❷

Volume of the smaller cuboid = $64\,cm^3 \div 8$

$= 8\,cm^3$

2 cm

4 cm

Exam tip

You should always reduce ratios to their simplest form when answering this type of question.

Key term

Ratio

Look out for

As you are finding the volume of the smaller shape, you need to divide by 8 to find the answer.

Exam-style questions

1. The two crosses in the diagram are mathematically similar. The area of the smaller shape is $90\,cm^2$.
   a Calculate the area of the larger shape. **[2]**
   The crosses are cross-sections of two mathematically similar prisms.
   b Write down the ratio of their volumes. **[1]**

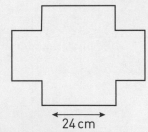

8 cm

24 cm

2. A manufacturer makes metal toy cars in two different sizes. The large car is an enlargement of the small car. The small car is 5 cm long and the large car 7.5 cm long. It takes $16\,cm^3$ of metal to make the small car.

   Calculate the volume of metal required to make the large car. **[3]**

3. The diagram shows two cones. Cone B is an enlargement of A. The radius of the base of cone A is 2 cm. The area of the base of cone B is 6 times larger than the area of cone A.

   Calculate the radius of the base of cone B. **[3]**

A

2 cm

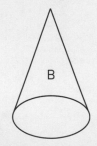

B

CHECKED ANSWERS

Constructing plans and elevations

Rules

1. Isometric drawings are used to accurately represent 3D objects
2. In Isometric drawings vertical edges are drawn vertically, horizontal edges are drawn slanted.
3. A plan of a 3D shape is the view from above.
4. An elevation of a 3D shape is the view from the front or side.

Key terms

Plan

Elevation

Isometric

Worked examples

a Make an isometric drawing from these elevations and plan. ① ②

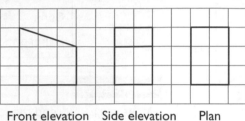

Front elevation   Side elevation   Plan

Answer

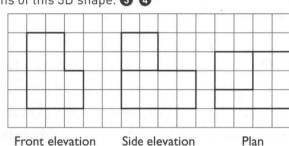

Front

Side

Look out for

Make sure the isometric paper is the right way up!

b Draw the plan, front and side elevations of this 3D shape. ③ ④

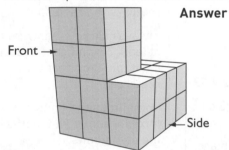

Front

Side

Answer

Front elevation   Side elevation   Plan

Exam tips

You should label the front and side of a 3D drawing.

You should clearly identify which diagram is the plan or elevation.

Exam-style questions

1. Make an isometric drawing of the shape that has these elevations and plan. **[3]**

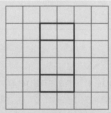

Front elevation   Side elevation   Plan

2. Draw the plan, front and side elevations of

a

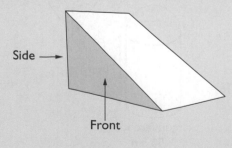

Side

Front

[3]

b

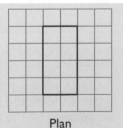

Side

Front   **[3]**

CHECKED ANSWERS

Surface area and 3D shapes

Rules

1. Volume of a pyramid $= \frac{1}{3}$ base area $\times$ height
2. Volume of a cone $= \frac{1}{3}\pi r^2 h$
3. Volume of a cylinder $= \pi r^2 h$
4. Surface area of a cylinder $= 2\pi rh + 2\pi r^2$
5. Volume of a sphere $= \frac{4}{3}\pi r^3$
6. Surface area of a sphere $= 4\pi r^2$

Key terms
Surface area
Pyramid
Cone
Sphere

Worked examples

a A cone has a radius of 4 cm and a height of 9 cm. Work out the volume of the cone. Give your answer in terms of π.

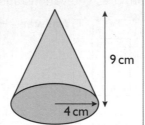

9 cm

4 cm

Answer

Volume of a cone $= \frac{1}{3}\pi r^2 h$ ②

Volume $= \frac{1}{3} \times \pi \times 4 \times 4 \times 9$

Volume $= 48\pi\,\text{cm}^3$

b A basket ball has a diameter of 24 cm. Calculate the volume and surface area of the basketball.

Answer

Volume $= \frac{4}{3}\pi r^3$ ⑤

Volume $= \frac{4}{3}\pi \times 12 \times 12 \times 12$

Volume $= 7230\,\text{cm}^3$ (3 s.f.)

Surface area $= 4\pi r^2$ ⑥

Surface area $= 4 \times \pi \times 12 \times 12$

Surface area $= 1810\,\text{cm}^2$ (3 s.f.)

c A wooden rod has a radius of 6 mm and is 10 cm long. Calculate the volume and surface area of the rod. Give your answer to 1 decimal place.

Answer

Volume of a cylinder $= \pi r^2 h$ ③

Volume $= \pi \times 0.6 \times 0.6 \times 10$

Volume $= 11.3\,\text{cm}^3$ (1 d.p.)

Surface area of a cylinder
$= 2\pi rl + 2\pi r^2$ ④

Surface area $= 2\pi \times 0.6 \times 10 +$
$2\pi \times 0.6 \times 0.6$

Surface area $= 40.0\,\text{cm}^2$ (1 d.p.)

Look out for
Make sure you read the question carefully, e.g. in this question you are told the diameter but the formulae requires the radius.

Always give your answers to at least 3 significant figures if not told the accuracy required.

Sometimes the units are mixed, change all lengths to the same units.

Exam tip
Students often get this type of question wrong because they do not change the measurements to the same units.

Exam-style questions

1 A cake mould is made in the shape of a square-based pyramid. The base of the mould has sides of 5 cm. It takes 50 cm³ of cake mixture to completely fill the mould. Work out the height of the cake mould. **[4]**

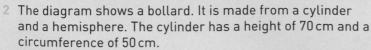

5 cm

2 The diagram shows a bollard. It is made from a cylinder and a hemisphere. The cylinder has a height of 70 cm and a circumference of 50 cm.
   a The bollard is going to be made of metal. Calculate the total volume of metal required to make the bollard. **[6]**
   b All exposed surfaces of the finished bollard will be painted white. Calculate the total surface area to be painted. **[4]**

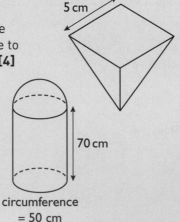

70 cm

circumference
= 50 cm

Area and volume in similar shapes

Rules

1. In general the ratio of lengths in similar figures is $a:b$.
2. In general the ratio of areas in similar figures is $a^2:b^2$.
3. In general the ratio of volumes or masses in similar figures is $a^3:b^3$.

Key terms

Ratio

Similar

Mass

Worked examples

a Two similar cones are made from card. It takes 125 cm² of card to make the larger cone. The larger cone has a height of 9 cm. The smaller cone has a height of 3 cm.
Calculate the area of card required to make the smaller cone.

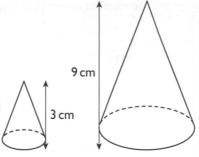

Answer

1 Ratio of height of cones is 9 : 3 or 3 : 1.

2 Ratio of surface areas of cones is $3^2 : 1^2$ or 9 : 1.
Amount of card required for smaller cone = 125 ÷ 9
= 13.9 cm²

b Two similar circular cake tins are to be used to bake two tiers of a wedding cake. The smaller tin has a radius of 8 cm and a height of 5 cm and requires 800 g of cake mixture. The larger tin requires 1.6 kg of cake mixture.
Calculate the radius and height of the larger tin.

Answer

Ratio of masses of cakes is 1 : 2.
Ratio of lengths in cake tins = $\sqrt[3]{1} : \sqrt[3]{2}$.
Radius of large tin is $8 \times \sqrt[3]{2}$ = 10.1 cm. ❸
Height of cake tin is $5 \times \sqrt[3]{2}$ = 6.3 cm (to nearest cm). ❸

Exam tip

Many students get this type of question wrong because the use the wrong scale factor or because they don't know whether to divide or multiply by the scale factor.

Look out for

Always use unrounded values of cube and square roots when calculating solutions to problems.

Exam-style questions

1 A furniture maker makes a set of two different sized coffee tables which are mathematically similar. The heights of the coffee tables are 40 cm and 50 cm. The area of the table top of the larger coffee table is 2475 cm².
Calculate the area of the smaller table top. **[3]**

2 A chocolate manufacturer makes chocolate bunnies in two sizes. The chocolate bunnies are mathematically similar. The smaller bunny is 11.3 cm high and has a mass of 100 g. The larger bunny has a mass of 500 g.
Calculate the height of the larger chocolate bunny. **[3]**

3 A manufacturer makes tins of beans in two sizes. The tins are mathematically similar cylindrical tins. The smaller tin contains 300 g and has a diameter of 6 cm and a height of 8 cm, The larger tin contains 500 g.
The tins have a label that covers the curved surface area of the tin.
Calculate the area of the label on the larger tin. **[5]**

CHECKED ANSWERS

Mixed exam-style questions

1 A ship sails on a bearing of 030° for 11.8 km.

Calculate exactly how far east the ship is from its starting point. [3]

2 A steel ball-bearing has a diameter of 8 mm.
The density of the steel the ball-bearing is made from is 7.8 g/cm³.
Calculate the mass of the ball bearing. [3]

3 The diameter of the Moon is 3474 km. The diameter of the Earth is 12 742 km.
Work out the ratio of the surface areas of the Moon and the Earth. Give your answer
in the form 1:x. [3]

4 A triangular prism has a cross-section in the shape of an equilateral triangle.
The sides of the triangle are 9 cm. The prism is 15 cm long.
a Calculate the area of the cross section. [4]
b Calculate the volume of the prism. [2]
There is a smaller version of the prism, mathematically similar to
the original. The length of the new prism is 5 cm.
c Calculate the area of the cross-section of the new prism. [2]
d Calculate the surface area of the new prism. [3]

5 Eleri is standing at 5 m from the foot of a cliff.
She sees the top of the cliff at an angle of elevation of 20°.
Eleri's eye is 1.4 m above the ground.

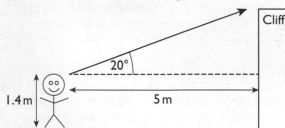

Work out the height of the cliff. [3]

6 A pyramid has a square base of 4 cm. The sloping edges of the pyramid are 6 cm long.
a Using ruler and pair of compasses only, make an accurate drawing of the
net of the pyramid. [4]
b Using measurements taken from your drawing, work out the surface
area of the pyramid. [4]

7 A water tank is made in the shape of a cylinder.
The cylinder is 1.3 m high and has a radius of 40 cm.
The tank is filled with water at the rate of 20 litres / minute.
Calculate the time taken to completely fill the water tank. [5]

8 Scale drawings of the plan, front and side elevations of a garden shed are shown below.

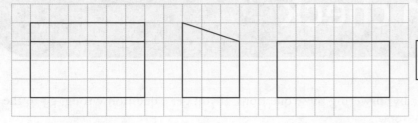

Key
1 square = 0.5 cm²

Front elevation   Side elevation   Plan

  a Make an isometric drawing of the shed.   [3]
The ratio of the scale drawing is 1:200.
The roof and four sides of the shed are to be painted in weatherproof paint.
The manufacturer of the paint states 1 litre will cover 12 m². The paint comes in 5 litre tins.
  b How many tins of paint will be required to cover the shed? Give a reason for your answer.   [5]

9 Describe fully the transformation that will map shape A onto shape B.   [2]

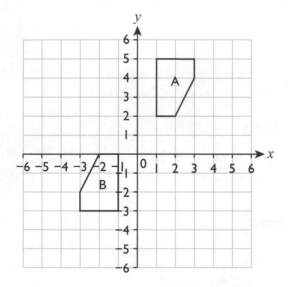

10 The diagram shows the design for a silver brooch.
AB and AC are tangents to the circle centre O, which has
a radius of 9 mm. BAC = 30°.
  a Calculate the area of the kite ABOC.   [5]
  b Calculate the area of the major sector of the circle.   [2]
The pendant is to be made of silver and will be 3 mm
thick. Silver has a density of 10.49 g/cm³.
  c Calculate the mass of the brooch to the nearest gram.   [4]

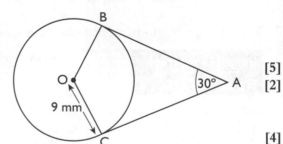

Statistics and Probability: pre-revision check

Check how well you know each topic by answering these questions. If you get a question wrong, go to the page number in brackets to revise that topic.

1 The box plots give information about the weights, in kg, of the fish in each of two lakes.

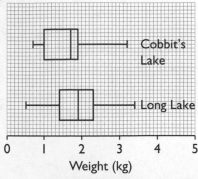

a Work out the inter-quartile range of the weights of the fish in Long Lake.
b Compare the distributions of the weights of the fish in these two lakes. (Page 89)

2 The table gives information about the age and the trunk radius of each of 8 trees.

Year (years)	26	42	50	33	55	58	36	48
Trunk radius (cm)	14	30	42	22	44	52	22	34

a Draw a frequency diagram to show this information.
b Describe and interpret the correlation shown in your frequency diagram.

Another tree has an age of 65 years.

c i Find an estimate for the trunk radius of this tree.
ii How reliable is your estimate? Explain why. (Page 91)

3 The table gives information about the heights, in metres, of some mountains.

Height (h metres)	$1000 < h \leqslant 2000$	$2000 < h \leqslant 2500$	$2500 < h \leqslant 3250$	$3250 < h \leqslant 4500$
Frequency	12	8	15	5

Draw a histogram to represent this data.   (Page 93)

4 There are 4 blue counters and 2 yellow counters in a bag.
Terri is going to take 2 counters at random from the bag.
Work out the probability that both counters will be the same colour. (Page 100)

5 In a survey of 50 people, 35 people said they like semi-skimmed milk, 40 people said they like skimmed milk and 28 people said they like both.
a Draw a Venn diagram to show this information.
b One of these people is taken at random. Work out the probability that this person does not like semi-skimmed milk or skimmed milk. (Page 102)

6 Here are some cards. Each card has a letter on it.

| S | T | A | T | I | S | T | I | C | S |

Naomi takes at random two of these cards. Work out the probability that she will take two Ts given that she will take at least one T. (Page 102)

Using grouped frequency tables

Rules

1. The modal group is the group with the highest frequency.
2. The median is the middle value when the data is in order of size.

   The middle value is the $\frac{n+1}{2}$ th value.

3. To calculate the mean from a grouped frequency table you need to add a mid-interval value column, a $f \times x$ column, and a total row to the table.
4. The estimated mean is the sum of all the $f \times x$ values divided by the number of values. Mean $= \frac{\Sigma f \times x}{n}$

Worked example

The table gives information about the heights of some children.

Height (x cm)	Frequency (f)	Mid-interval value	$f \times x$
$130 < x \leqslant 140$	10	135	$10 \times 135 = 1350$
$140 < x \leqslant 150$	16	145	$16 \times 145 = 2320$
$150 < x \leqslant 160$	17	155	$17 \times 155 = 2635$
$160 < x \leqslant 170$	7	165	$7 \times 165 = 1155$
Total	**50**		7460

Key terms

Frequency

Mid-interval value

Exam tip

Give the units with your answer.

i Write down the modal group.

ii Find the group that contains the median height.

iii Work out an estimate for the mean height.

Answer

i The modal group is $150 < x \leqslant 160$ ❶. The group with the highest frequency (17) is $150 < x \leqslant 160$, so this is the modal group.

ii The group that contains the median height is $140 < x \leqslant 150$ ❷. The middle value of the data is the $\frac{50+1}{2}$ th = 25.5th value, i.e. half way between the 25th and 26th values. There are 10 values in the group $130 < x \leqslant 140$ and 16 values in the group $140 < x \leqslant 150$, so both the 25th and 26th values appear in the group $140 < x \leqslant 150$, so the median value is in the group $140 < x \leqslant 150$.

iii Add a mid-interval value column, a $f \times x$ column, and a total row to the table ❸. Mean $= \frac{7460}{50} =$ 149.2 cm ❹. The sum of all the $f \times x$ values is 7460, the number of values is 50, so the estimated mean is $7460 \div 50 = 149.2$

Exam-style questions

1 The table shows information about the time taken, in seconds, for each of 70 people to complete a logic problem.

Time taken (x seconds)	Frequency (f)
$30 < x \leqslant 40$	5
$40 < x \leqslant 50$	8
$50 < x \leqslant 60$	12
$60 < x \leqslant 70$	28
$70 < x \leqslant 80$	17

Work out an estimate for the mean time taken. **[4]**

2 Jai recorded the weights, in kg, of some packages. His results are summarised in the table.

Weight (w kg)	Frequency (f)
$1 < w \leqslant 1.5$	29
$1.5 < w \leqslant 2$	17
$2 < w \leqslant 2.5$	11
$2.5 < w \leqslant 3$	8

a Write down the modal group. **[1]**

b Find the group that contains the median weight. **[1]**

c Work out an estimate for the mean weight. **[4]**

Inter-quartile range

Rules

1 For continuous data, lower quartile = $\frac{n}{4}$ th value of the data, median = $\frac{n}{2}$ th value, upper quartile = $\frac{3n}{4}$ th value of the data.

2 For discrete data, lower quartile = $\frac{n+1}{4}$ th value of the ordered data, median = $\frac{n+1}{2}$ th value, upper quartile = $\frac{3(n+1)}{4}$ th value of the ordered data.

3 Inter-quartile range = upper quartile – lower quartile.

Key terms

Continuous data

Discrete data

Worked examples

a The cumulative frequency diagram gives information about the weights of 48 samples of moon rock.

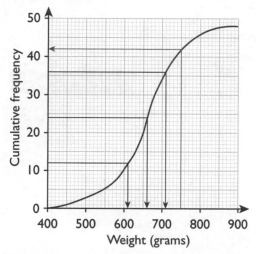

i Find an estimate of the median weight.

ii Find an estimate of the inter-quartile range.

iii Work out an estimate for the percentage of these samples that have a weight greater than 750 grams.

Answer

i There are 48 samples of moon rock. Weight is continuous data.

So the median weight is the $\frac{48}{2}$ th = 24th value = 665 grams from cumulative frequency diagram. **1**

ii The lower quartile = $\frac{48}{4}$ th = 12th value = 610 grams from cumulative frequency diagram. **1**

The upper quartile = $\frac{3 \times 48}{4}$ th = 36th value = 710 grams from cumulative frequency diagram. **1**

So the inter-quartile range = 710 – 610 = 100 grams. **3**

iii There are 6 samples of moon rook with a weight greater than 750 grams, i.e. 48 – 42 = 6 from cumulative frequency diagram.

So the percentage of samples with a weight greater than 750 grams = $\frac{6}{48} \times 100 = 12.5\%$.

Exam tips

Show your working by drawing appropriate lines on cumulative frequency diagrams.

Exam-style questions

1 The table gives information about the average temperature in Newcastle on each of 40 days.
   a   Draw a cumulative frequency diagram for this information. **[4]**
   b   Find an estimate for the inter-quartile range of the temperatures. **[2]**

Average temperature (A °C)	$15 < A \leq 17$	$17 < A \leq 19$	$19 < A \leq 21$	$21 < A \leq 23$
Frequency	3	14	20	3

2 The box plots show information about times taken, in minutes, for each of 100 journeys by road in 1995 and for the same journeys by road in 2015.

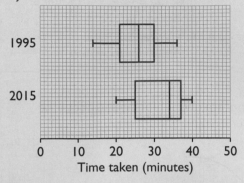

Compare these distributions. **[2]**

CHECKED ANSWERS

Displaying grouped data

Rules

1. Use groups of equal width when drawing a frequency diagram.
2. Tallies are used to record data in the appropriate groups.
3. Complete the frequency column in the grouped frequency table by totalling tallies.
4. A jagged line is used to show that the scale on an axis does not start at zero.

Worked examples

24 students entered a high jump competition. Here is the best height, h metres, jumped by each student.

1.1	1.4	1.3	1.3	1.6	1.1	1.5	1.3	1.1	1.4	1.2	1.5
1.5	1.3	1.6	1.2	1.3	1.8	1.7	1.3	1.7	1.5	1.9	1.3

i   Display the information in a grouped frequency table. Use the groups $1 < h \leqslant 1.2$, $1.2 < h \leqslant 1.4$, and so on.

ii Draw a frequency diagram to show the data.

iii Describe the shape of the distribution.

Answer

i

Height (h metres)	Tally	Frequency
$1 < h \leqslant 1.2$	卌	5
$1.2 < h \leqslant 1.4$	卌 IIII	9
$1.4 < h \leqslant 1.6$	卌 I	6
$1.6 < h \leqslant 1.8$	III	3
$1.8 < h \leqslant 2$	I	1

Draw a grouped frequency table for the information. Continue the pattern of the groups. The width of each group is equal to 0.2 metres. ❶

Use tallies to complete the grouped frequency table. ❷

Complete the frequency column in the grouped frequency table. ❸

Draw a frequency diagram for the information in your table. Use a jagged line to show that the scale on the horizontal axis does not start at zero. ❹

ii

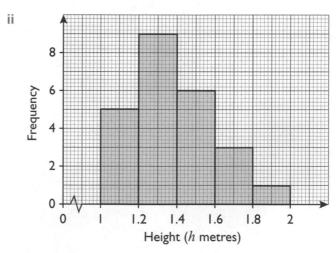

iii The modal group is $1.2 < h \leqslant 1.4$. The distribution is not symmetrical. The modal group is on the left of the distribution so it is positively skewed.

Key terms

Discrete data

Continuous data

Skew

Class interval

Exam-style questions

Franz recorded the time taken, t seconds, for each of 25 telephone calls. Here are his results.

20.6	5.7	20.1	11.2	25.8	13.7	26.8	27.9	14.6	24.3	21.7	25.2	18.1
16.9	24.6	22.8	21.9	19.6	26.7	23.7	18.4	17.0	28.4	29.5	22.3	

a   Time taken is an example of continuous data. Explain why. **[1]**

b   Display the information in a grouped frequency table. Use the groups $5 < t \leq 10$, $10 < t \leq 15$, and so on. **[3]**

c   Draw a frequency diagram to show the data. **[3]**

d   Franz says that the group that contains the median time taken is the same group as the modal group. Is he right? Explain why. **[2]**

CHECKED ANSWERS

Histograms

Rules

1. Frequency density = $\frac{\text{frequency}}{\text{class width}}$.
2. Frequency = frequency density × class width.

Worked examples

a The incomplete table (blue) and histogram (blue) give some information about the iron content of a sample of rocks.

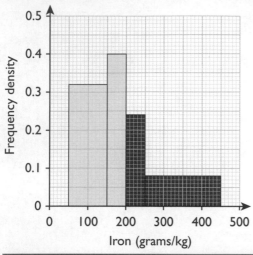

Iron (x grams/kg)	Frequency	Frequency density
$50 < x \leqslant 150$	32	0.32
$150 < x \leqslant 200$	20	0.4
$200 < x \leqslant 250$	12	0.24
$250 < x \leqslant 450$	16	0.08

i Use the table to complete the histogram.
ii Use the histogram to complete the table.

Answers

i For the class $200 < x \leqslant 250$, class width = 250 − 200 = 50 and frequency = 12 (from table).
So frequency density = $\frac{\text{frequency}}{\text{class width}} = \frac{12}{50} = 0.24$. ❶

For the class $250 < x \leqslant 450$, class width = 450 − 250 = 200 and frequency = 16 (from table).

So frequency density = $\frac{\text{frequency}}{\text{class width}} = \frac{16}{200} = 0.08$. ❶

Complete the histogram (shown in red on the histogram above).

ii For the class $50 < x \leqslant 150$, class width = 150 − 50 = 100 and frequency density =0.32 (from histogram).
So frequency = frequency density × class width = 0.32 × 100 = 32. ❷
For the class $150 < x \leqslant 200$, class width = 200 − 150 = 50 and frequency density = 0.4 (from histogram).
So frequency = frequency density × class width = 0.4 × 50 = 20. ❷
Complete the table (shown in blue in the table above).

Exam tip

Add a column to the grouped frequency table and record the frequency densities.

1 Jamie recorded the weights of 100 loaves of bread. His results are summarised in the table below.

Weight (w grams)	Frequency
$450 < w \leqslant 480$	15
$480 < w \leqslant 500$	25
$500 < w \leqslant 510$	24
$510 < w \leqslant 550$	36

Draw a histogram to show this information. **[4]**

2 The histogram shows information about the heights of some trees.

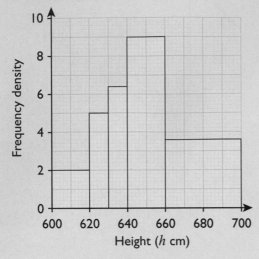

a Work out the number of trees with a height less than 620 cm. **[2]**
b Work out an estimate for the median height. **[4]**

CHECKED ANSWERS

Mixed exam-style questions

1 John recorded the times taken, in minutes, for each of eight students to complete a 250-piece jigsaw puzzle and a 500-piece jigsaw puzzle. His results are given in the table.

250-piece jigsaw (minutes)	35	41	70	71	62	74	45	51
500-piece jigsaw (minutes)	68	70	70	90	86	99	75	78

   a Draw a scatter diagram for this information.   **[3]**

   b One of the data points may be an outlier. Which data point? Give a reason for your answer.   **[1]**

   c Describe and interpret the correlation.   **[2]**

   Kyle is another student. It takes him 57 minutes to do the 250-piece jigsaw.

   d i   Find an estimate for how long it takes him to do the 500-piece jigsaw.

      ii Comment on the reliability of your estimate.   **[3]**

2 The box plot shows information about the recorded mileages of some cars.

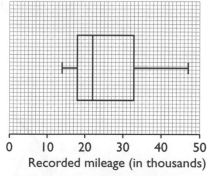

Recorded mileage (in thousands)

   a Find the median recorded mileage.   **[1]**

   b Work out the inter-quartile range.   **[2]**

   c What percentage of these cars have a recorded mileage greater than 33 000?   **[2]**

3 Pym recorded the lengths of time, t seconds, that each of 120 dancers could stand on one leg. The incomplete histogram shows some information about his results. Pym recorded 22 dancers in the class $0 < t \leqslant 100$.

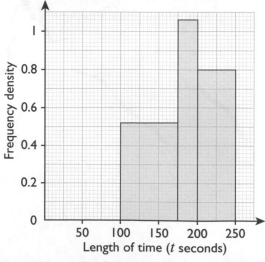

   a Work out the frequency density for this class.   **[2]**

   b Work out an estimate for the number of dancers who could stand on one leg for more than 2 minutes.   **[3]**

4 The table gives information about the absolute magnitude of the intensity of light coming from each of 46 stars in the constellation Ursa Minor.

Absolute magnitude (M)	$-4 < M \leqslant -1$	$-1 < M \leqslant 0$	$0 < M \leqslant 1$	$1 < M \leqslant 3$	$3 < M \leqslant 6$
Frequency	10	10	10	8	8

Draw a histogram to show this information. [4]

5 40 people bought some fruit in a shop.
Of these, 23 people bought apples and 18 people bought bananas.
Some people bought both apples and bananas and 10 people bought neither.
One of these people is picked at random.
Given that this person bought apples, find the probability that they also bought bananas. [4]

6 Box A contains 3 green buttons and 2 orange buttons.
Box B contains 2 green buttons and 5 orange buttons.
Box C contains 4 green buttons and 7 orange buttons.
Hattie takes at random a button from box A.
If the button is green she will take a button at random from box B.
If the button is orange she will take a button at random from box C.
Find the probability that the second button will be green given that she takes two buttons of the same colour. [6]

Working with stratified sample techniques and defining a random sample

REVISED ☐

HIGH

Rules

1. A random sample means every data item has an equal chance of being selected.
2. A stratified sample means that the sample selected should be in proportion to the data that has been given.
3. With a stratified sample always work out for each category, as there may be a need to look back at your working. Do not leave out any working.
4. Never round until the last stage of working, after each category has been worked out.

Worked examples

a Is selecting all students with the surname beginning with the letter W to answer a questionnaire a method of using a random sample? Give a reason for your answer.

Answer

No. As not all students have an equal chance of being selected. For example, if your surname begins with an H you have no chance of being selected.

b A conference with 50 people attending is to be a stratified sample selected from employees working in three different countries.

How many employees should attend from each of the three countries?

Country	Number of employees
Italy	2340
Poland	7725
Sweden	9230

Answer

Total number of employees 2340 + 7725 + 9230 = 19295.

Proportion from each country:

$$\frac{\text{Number from that country}}{\text{Total number of employees}} \times \text{number of people attending}$$

Give answers to a few decimal places to start, then round to a whole number of people later.

Italy $\frac{2340}{19\,295} \times 50 = 6.0637...$ Poland $\frac{7725}{19\,295} \times 50 = 20.0181...$ Sweden $\frac{9230}{19\,295} \times 50 = 23.918...$

After rounding answers to a whole number of people, you must check that the total is the 50 required.

6 + 20 + 24 = 50. So: 6 employees from Italy, 20 from Poland and 24 from Sweden.

Key terms

Random

Selection

Sample

Stratified

Exam-style questions

1 Here is the population of three villages. A council with a total of 16 members with representatives from all three villages is to be formed. A stratified sampling method is to be used. How many members will there be from each of the three villages on the council?

Village	Population
Brewyn	3050
Dafddu	6735
Cae Ben	5634

CHECKED ANSWERS ☐

Exam tips

If your total when working with a stratified sample does not add to the number you expect, for example, you may be 1 or 2 too many, then go back to adjust your decisions based on a few decimal places.

Always work out all the calculations as a proportion, then check back on your rounding.

The multiplication rule

Rules

1. $P(A) + P(\text{not } A) = 1$
2. For independent events, $P(A \text{ and } B) = P(A) \times P(B)$
3. For mutually exclusive events, $P(A \text{ or } B) = P(A) + P(B)$

Worked example

There are three blue crayons and two red crayons in a box. Tina takes at random two crayons from the box.

   i Draw a tree diagram to show this situation.

  ii Work out the probability that both crayons will be the same colour.

Answers

   i Draw the tree diagram.

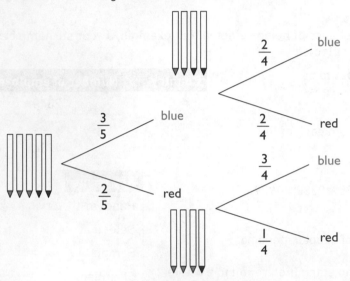

Take the crayons one at a time from the box. The colour of the first crayon affects the crayons left in the box. If the first crayon is blue there will be two blue crayons and two red crayons left in the box. If the first crayon is red, there will be three blue crayons and one red crayon left in the box.

The probabilities on each pair of branches must add to 1, as $P(\text{blue}) + P(\text{not blue, i.e. red}) = 1$ ❶

Key terms

Independent events

Dependent events

Mutually exclusive events

  ii For both crayons to be the same colour either two blue crayons are taken from the box or two red crayons are taken from the box.

The probability of taking a blue crayon first and a blue crayon second is:

$P(\textbf{blue first and blue second}) = P(\text{blue first}) \times P(\text{blue second}) = \frac{3}{5} \times \frac{2}{4} = \frac{6}{20}$ ❷

The probability of taking a red crayon first and a red crayon second is:

$P(\textbf{red first and red second}) = P(\text{red first}) \times P(\text{red second}) = \frac{2}{5} \times \frac{1}{4} = \frac{2}{20}$ ❷

So, the probability of taking two blue crayons or two red crayons is:

$P(\textbf{blue first and blue second or red first and red second}) =$

$P(\textbf{blue first and blue second}) + P(\textbf{red first and red second}) = \frac{6}{20} + \frac{2}{20} = \frac{8}{20} = \frac{2}{5}$ ❸

Exam hints

Multiply the probabilities 'along' the branches of the tree diagram.

Add the probabilities 'down' the branches of the tree diagram.

Leave the fractions unsimplified, including the final answer.

Exam-style questions

1 Taavi is given a raffle ticket on Wednesday and a raffle ticket on Saturday. The probability that he will win on Wednesday is 0.1. The probability that he will win on Saturday is 0.05.

a Draw a tree diagram to show this information. **[3]**

b Work out the probability that he will win on both days. **[2]**

c Work out the probability that he will not win on Wednesday and win on Saturday. **[2]**

2 A bag contains three green counters and four yellow counters. A box contains one green counter and five yellow counters. Jim takes at random a counter from the box. If the counter is green Jim takes at random a counter from the bag. If the counter is yellow Jim takes another counter from the box. Find the probability that the colour of the first counter will be different to the colour of the second counter. **[5]**

CHECKED ANSWERS

The addition rule and Venn diagram notation

LOW

Rules

1. P(event happening) = $\dfrac{\text{total number of successful outcomes}}{\text{total number of possible outcomes}}$
2. For mutually exclusive events, P(A or B) = P(A) + P(B)
3. For events that are not mutually exclusive, P(A or B) = P(A) + P(B) – P(A and B)

Worked examples

a A box contains 17 counters, of which 6 counters are black and 3 counters are white. The Venn diagram shows this information. A counter is taken at random from the box. Find the probability the counter will be
  i black
  ii white
  iii black or white.

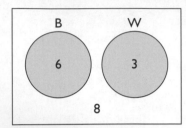

Answers

a i P(B) = $\dfrac{\text{total number of successful outcomes}}{\text{total number of possible outcomes}} = \dfrac{6}{17}$ **1**

ii P(W) = $\dfrac{\text{total number of successful outcomes}}{\text{total number of possible outcomes}} = \dfrac{3}{17}$ **1**

iii Taking a black counter and taking a white counter are mutually exclusive, they cannot both happen,

so P(B or W) = P(B) + P(W); P(B or W) = $\dfrac{6}{17} + \dfrac{3}{17} = \dfrac{9}{17}$ **2**

Key terms

Mutually exclusive events

Independent events

b In a survey 29 students were asked if they like celery or rhubarb. 12 said they like celery, 13 said they like rhubarb and 11 said they like neither celery nor rhubarb. The Venn diagram shows this information. One of these students is picked at random. Find the probability that this student likes
  i celery
  ii rhubarb
  iii celery and rhubarb
  iv celery or rhubarb.

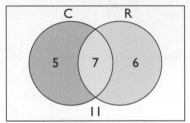

Answers

b i P(C) = $\dfrac{\text{total number of successful outcomes}}{\text{total number of possible outcomes}} = \dfrac{12}{29}$ **1**

ii P(R) = $\dfrac{\text{total number of successful outcomes}}{\text{total number of possible outcomes}} = \dfrac{13}{29}$ **1**

iii 7 students like celery and rhubarb.

So P(C and R) = $\dfrac{\text{total number of successful outcomes}}{\text{total number of possible outcomes}} = \dfrac{7}{29}$ **1**

iv Liking celery and liking rhubarb are not mutually exclusive, 7 students like both, so P(C or R) = P(C) + P(R) – P(C and R);
P(C or R) = $\dfrac{12}{29} + \dfrac{13}{29} - \dfrac{7}{29} = \dfrac{18}{29}$

Exam tips

Write down the appropriate formula before using it.

Draw a Venn diagram to show all the information.

c Shade the following regions on this Venn diagram.

  i A ∪ B ∪ C

  ii A'

  iii B ∩ C

Answer

  i ∪ means union, 'together with'. So, A together with B together with C.

  ii A' means the complement of A. So, every region except A.

  iii B ∩ C means the intersection (∩) of B and C.

i 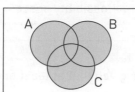   **ii** 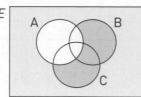   **iii**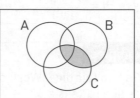

Exam-style questions

1 The Venn diagram shows information about the numbers and colours of beads in a bag.

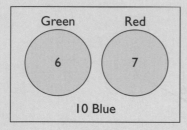

A bead is taken at random from the bag. Find the probability that the bead will be

  a green

  b green or blue

  c green and red. **[4]**

2 67 people go to a concert. 35 of these people buy a programme, 27 buy an ice cream and 18 buy neither a programme nor an ice cream. Some people buy both.

  a Draw a Venn diagram to show this information. **[3]**

  b One of these people is picked at random. Find the probability that this person buys

    i only a programme

    ii a programme or an ice cream. **[3]**

3 A and B are two independent events: P(A) = 0.8, P(B) = 0.5

  a Find P(A and B) **[2]**

  b Find P(A or B) **[2]**

CHECKED ANSWERS

Conditional probability

Rules

❶ P(event happening) = $\frac{\text{total number of successful outcomes}}{\text{total number of possible outcomes}}$.

❷ For independent events, P(A and B) = P(A) × P(B).

❸ P(A) = 1 – P(not A).

❹ P(A given B) = $\frac{\text{P(A and B)}}{\text{P(B)}}$.

Key terms

Independent event

Conditional probability

Possibility space

Worked examples

a Yani rolls two fair dice. Work out the probability that she rolls a double 6 given that at least one of the numbers she rolls is a 6.

Answers

Method 1

List the relevant outcomes or draw a possibility space diagram showing all the possible outcomes.

There are 11 possible outcomes given that Yani rolls at least one 6: (1, 6), (2, 6), (3, 6), (4, 6), (5, 6), (6, 6), (6, 5), (6, 4), (6, 3), (6, 2), (6,1).

Of these, only 1 of these is a successful outcome: (6, 6).

So P(double 6) = $\frac{\text{total number of successful outcomes}}{\text{total number of possible outcomes}} = \frac{1}{11}$. ❶

Method 2 (useful in more complicated problems)

Draw a tree diagram for the information.

P(double 6) = $\frac{1}{6} \times \frac{1}{6} = \frac{1}{36}$ ❷

P(at least one 6) = 1 – P(no 6s) ❸

$= 1 - \frac{5}{6} \times \frac{5}{6} = 1 - \frac{25}{36}$

$= \frac{11}{36}$

So P(double 6 given at least one 6)

$= \frac{\text{P(double 6 and at least one 6)}}{\text{P(at least one 6)}} = \frac{\left(\frac{1}{36}\right)}{\left(\frac{11}{36}\right)} = \frac{1}{11}$ ❹

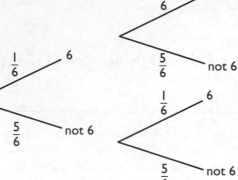

Exam tip

Show your working by writing down the formulas you are using.

Exam-style questions

1 The Venn diagram gives information about the numbers of people who drive (D) to work or who cycle (C) to work. Some do both and some do neither. One of these people is picked at random.
   a Find the probability that this person:
     i cycles
     ii drives and cycles to work. **[2]**
   b Find the probability that this person drives to work given that they cycle to work. **[2]**

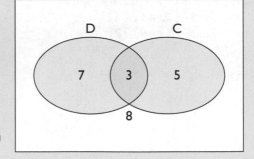

2 Michael spins three coins. Find the probability that he will get two Heads given that he gets at least one Tail. **[4]**

3 A bag contains 4 red beads and 7 green beads. Kieran takes at random two beads from the bag.

   Find the probability that both beads are green given that both beads are the same colour. **[5]**

CHECKED ANSWERS ☐

Mixed exam-style questions

1 An insurance company received a total of 3467 claims last year. Of these 2125 were for household damage. This year the insurance company expects to receive a total of 5000 claims. Estimate the number of claims for household damage this year. **[2]**

2 Cheri spins a fair 5–sided spinner, numbered 1, 2, 3, 4 and 5, and rolls an ordinary dice. What is the probability that Cheri will get
   a a 3 on the spinner and a 3 on the dice **[2]**
   b a 3 on the spinner or a 3 on the dice? **[2]**

3 Bag A contains 3 red counters and 2 green counters. Bag B contains 4 red counters and 5 green counters. Bag C contains 1 red counter and 6 green counters. Hamish is going to take at random a counter from bag A. If he takes a red counter he will take a counter at random from bag B. If he takes a green counter he will take a counter at random from bag C.
   Work out the probability that both counters will be the same colour. **[4]**

4 Nima rolls 5 ordinary dice. Work out the probability that he will get exactly three 6s. **[3]**

5 24 people gave presents on Mothers' day. 7 people gave only flowers. 5 people gave only chocolates, x people gave both flowers and chocolates, and 9 people gave neither flowers nor chocolates.
   a Draw a Venn diagram to show this information. **[3]**
   b Work out the value of x. **[2]**
   c One of these people is picked at random. Find the probability that this person gave
      i both flowers and chocolates
      ii flowers or chocolates or both. **[3]**

The language used in mathematics examinations

- **You must show your working...** you will lose marks if working is not shown.
- **Estimate...** often means round numbers to 1 s.f.
- **Calculate...** some working out is needed; so show it!
- **Work out / find...** a written or mental calculation is needed.
- **Write down...** written working out is not usually required.
- **Give an exact value of...** no rounding or approximations:
  - on a calculator paper write down all the numbers on your calculator
  - on a non-calculator paper give your answer in terms of π, a fraction or a surd.
- **Give your answer to an appropriate degree of accuracy...** if the numbers in the question are given to 2 d.p. give your answer to 2 d.p.
- **Give your answer in its simplest form...** usually cancelling of a fraction or a ratio is required.
- **Simplify...** collect like terms together in an algebraic expression.
- **Solve...** usually means find the value of x in an equation.
- **Expand...** multiply out brackets.
- **Construct, using ruler and compasses...** the ruler is to be used as a straight edge and compasses must be used to draw arcs. You **must** show all your construction lines.
- **Measure...** use a ruler or a protractor to accurately measure lengths or angles.
- **Draw an accurate diagram...** use a ruler and protractor – lengths must be exact, angles must be accurate.
- **Make y the subject of the formula...** rearrange the formula to get y on its own on one side e.g. $y = \frac{2x - 3}{4}$.
- **Sketch...** an accurate drawing is not required – freehand drawing will be accepted.
- **Diagram NOT accurately drawn...** don't measure angles or sides from the diagram – you must calculate them if you are asked for them.
- **Give reasons for your answer... OR explain why...** worded explanations are required referring to the theory used.
- **Use your/the graph...** read the values from your graph and use them.
- **Describe fully...** usually transformations:
  - Translation
  - Reflection in a line
  - Rotation through an angle about a point
  - Enlargement by a scale factor about a point
- **Give a reason for your answer...** usually in angle questions, a written reason is required e.g. 'angles in a triangle add up to 180°' or 'corresponding angles are equal', etc.
- **You must explain your answer...** a worded explanation is required along with the answer.
- **Show how you got your answer...** you must show all your working, words may also be needed.
- **Describe...** answer the question using words.
- **Write down any assumption you make...** describe any things you have assumed are true when giving your answer.
- **Show...** usually requires you to use algebra or reasons to show something is true.
- **Complete...** Finish off a table or a diagram
- **Units...** Write down the units that goes with the question e.g. m per sec or ms^{-1}

Exam technique and formulae that will be given

- Be prepared and know what to expect.
- Don't just learn key points.
- Work through past papers. Start from the back and work towards the easier questions. Your teacher will be able to help you.
- Practice is the key, it won't just happen.
- Read the question thoroughly.
- Cross out answers if you change them, only give **one** answer.
- Underline the key facts in the question.
- Estimate the answer.
- Is the answer right/realistic?
- Have the right equipment.
  - ○ Calculator
  - ○ Pens
  - ○ Pencils
  - ○ Ruler, compass, protractor
  - ○ Eraser
  - ○ Tracing paper
  - ○ Spares
- Never give two different answers to a question.
- Never just give just an answer if there is more than 1 mark.
- Never measure diagrams; most diagrams are not drawn accurately.
- Never just give the rounded answer; always show the full answer in the working space.
- Read each question carefully.
- Show stages in your working.
- Check your answer has the units.
- Work steadily through the paper.
- Skip questions you cannot do and then go back to them if time allows.
- Use marks as a guide for time: 1 mark = 1 min
- Present clear answers at the bottom of the space provided.
- Go back to questions you did not do.
- Read the information below the diagram – this is accurate.
- Use mnemonics to help remember formula you will need, for example:
  - ○ SOH   sin = opposite/hypotenuse
  - ○ CAH   cos = adjacent/hypotenuse
  - ○ TOA   tan = opposite/adjacent
  - ○ or 'silly old hens cackle and hale, till old age'.
  - ○ For the order of operations, BIDMAS: Brackets, Indices, Division, Multiply, Add, Subtract
  - ○ Formula triangles for the relationship between three parameters e.g. speed, distance and time

Distance = Speed × Time

Time = $\dfrac{\text{Distance}}{\text{Speed}}$

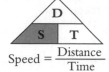
Speed = $\dfrac{\text{Distance}}{\text{Time}}$

- The following formulae will be provided for students within the relevant examination questions. All other formulae and rules **must** be learnt. Where r is the radius of the sphere or cone, l is the slant height of a cone and h is the perpendicular height of a cone:
  - ○ Curved surface area of a cone = $\pi r l$
  - ○ Volume of a cone = $\frac{1}{3}\pi r^2 h$
  - ○ Surface area of a sphere = $4\pi r^2$
  - ○ Volume of a sphere = $\frac{4}{3}\pi r^3$

Common areas where students make mistakes

Here are some topics that students frequently make errors in during their exam.

Number

	Question	Working	Answer
Estimating	Estimate $\dfrac{76.15 \times 0.49}{19.04}$	Write each number to one significant figure so that: 76.15 becomes 80 0.49 becomes 0.5 19.04 becomes 20 Remember that the size of the estimate needs to be similar to the original number. So $80 \times 0.5 = 40$ and $40 \div 20 = 2$	2

	Question	Working	Answer
Using a calculator	Work out $\dfrac{76.15 + 5.62^2}{19.04}$	You either need to enter the whole calculation into your calculator using the fraction button or work out the top first then divide the answer by the bottom. $76.15 + 5.62^2 = 107.7344$ $107.73 \div 19.04 = 5.658319327$	5.658319327

Fractions	Question	Working	Answer
Adding	$5\frac{2}{3} + 2\frac{1}{4}$	Deal with the whole numbers first $5 + 2 = 7$ then with the fractions by writing them as equivalent fractions $\frac{2}{3} = \frac{4}{6} = \frac{6}{9} = \frac{8}{12}$ and $\frac{1}{4} = \frac{2}{8} = \frac{3}{12}$ 12 is the LCM of 3 and 4 so write the fractions in 12ths. $\frac{8}{12} + \frac{3}{12} = \frac{8+3}{12} = \frac{11}{12}$	$7\frac{11}{12}$
Subtracting	$5\frac{2}{3} - 2\frac{1}{4}$	You use the same method as adding but just subtract so we get $5 - 2 = 3$ and $\frac{8}{12} - \frac{3}{12} = \frac{8-3}{12} = \frac{5}{12}$	$3\frac{5}{12}$
Multiplying	$3\frac{2}{5} \times 2\frac{3}{4}$	Write the fractions as improper fractions then multiply the tops together and then the bottoms of the fractions. $\frac{17}{5} \times \frac{11}{8} = \frac{187}{40}$ then cancel by dividing by 40.	$4\frac{27}{40}$
Dividing	$3\frac{1}{2} \div 2\frac{2}{3}$	Write the fractions as improper fractions, write the first fraction down and turn the second fraction upside down and multiply $\frac{7}{2} \div \frac{8}{3} = \frac{7}{2} \times \frac{3}{8} = \frac{21}{16}$ Then write as a mixed number.	$1\frac{5}{16}$

	Question	Working	Answer
Finding a reverse percentage	Find original price if the sale price is £60 after a 20% deduction.	80% of the original price is £60 100% of original price is $60 \div 80 \times 100$	£75

	Question	Working	Answer
Working with bounds	Find the upper bound of petrol consumption if 238 miles are driven to nearest mile, using 27.3 litres of petrol to nearest tenth of a litre.	Maximum value obtained by: upper bound of miles ÷ lower bound of petrol used = $238.5 \div 27.25 =$	8.75 correct to 3 s.f.

Algebra

Rules

$$n^a \times n^b = n^{a+b} \quad n^a \div n^b = n^{a-b} \quad \left(n^a\right)^b = n^{a \times b} \quad n^{\frac{1}{2}} = \sqrt{n} \quad n^{\frac{1}{3}} = \sqrt[3]{n}$$

$$n^{-a} = \frac{1}{n^a} \quad \sqrt{a \times b} = \sqrt{a} \times \sqrt{b} \quad \sqrt{\frac{a}{b}} = \frac{\sqrt{a}}{\sqrt{b}} \quad (a-b)(a+b) = a^2 - b^2$$

	Question	Working	Answer
Index laws	Simplify **a** $7f^4 g^3 \times 2f^3 g$ **b** $\dfrac{12t^5}{u^4} \times \dfrac{u^3}{3t^2}$ **c** $(y^2)^3$ **d** $\left(9x^2 y^4\right)^{\frac{3}{2}}$	$= 7 \times 2 \times f^{4+3} \times g^{3+1}$ $= \dfrac{12}{3} t^{5-2} u^{3-4} = 4t^3 u^{-1}$ $= y^{2 \times 3}$ $= 9^{\frac{3}{2}} x^{2 \times \frac{3}{2}} y^{4 \times \frac{3}{2}} = 27x^3 y^6$	$14f^7 g^4$ $\dfrac{4t^3}{u}$ y^6 $27x^3 y^6$

	Question	Working	Answer
Multiplying out brackets	Expand **a** $7p - 4(p - q)$ **b** $(y + 3)(y - 4)$ **c** $(3a + 2b)(3a - 2b)$	$= 7p - 4 \times p - 4 \times -q = 7p - 4p + 4q$ $= y \times y + y \times -4 + 3 \times y + 3 \times -4$ $= y^2 - 4y + 3y - 12$ $= 3a \times 3a + 3a \times -2b + 2b \times 3a +$ $2b \times -2b = 9a^2 + 6ab - 6ab - 4b^2$	$3p + 4q$ $y^2 - y - 12$ $9a^2 - 4b^2$

	Question	Working	Answer
Factorising expressions	Factorise completely **a** $12e^2 f - 9ef^2$ **b** $x^2 - 7x + 12$ **c** $6x^2 - 11x + 3$ **d** $25p^2 - 9t^2$	$= 3 \times 4 \times e \times e \times f - 3 \times 3 \times e \times f \times f$ $= x^2 - (3 + 4)x + -3 \times -4$ $= 2 \times 3x^2 - (5 + 6)x + -3 \times -1$ Difference of two squares so =	$3ef(4e - 3f)$ $(x - 3)(x - 4)$ $(2x - 3)(3x - 1)$ $(5p + 3t)(5p - 3t)$

	Question	Working	Answer
Solving equations	Solve		
	a $3f + 4 = 5f - 3$	$4 + 3 = 5f - 3f$ so $7 = 2f$ or $2f = 7$	$f = 3.5$
	b $5(x + 2) = 3$	$5x + 10 = 3$ so $5x = 3 - 10$ or $5x = -7$	$x = -1.4$
	c $y^2 - 3y - 10 = 0$	$(y + 2)(y - 5) = 0$ so $y + 2 = 0$ or $y - 5 = 0$	$y = -2$ or $y = 5$
	d $2x + 3y = 7$	$2 \times (2x + 3y = 7)$ $\quad = 4x + 6y = 14$	
	$\quad 3x - 2y = 17$	$3 \times (3x - 2y = 17)$ $\quad = 9x - 6y = 51$	
		Adding eliminates ys so $13x = 65$	
		So $x = 5$; substituting into $2x + 3y = 7$	
		gives $2 \times 5 + 3y = 7$ so $y = -1$	$x = 5, y = -1$

	nth term	Notes	Series is
Linear series	$a + (n - 1)d$	a is the first term and d the difference between each term	$a, a + d, a + 2d, \ldots$

	Question	Working	Answer
Simplifying fractions	Write as a single fraction $3 + \dfrac{6x + 15}{2x^2 + 7x + 5}$	Use common denominator. $\dfrac{3(2x^2 + 7x + 5)}{2x^2 + 7x + 5} + \dfrac{6x + 15}{2x^2 + 7x + 5}$ $\dfrac{6x^2 + 21x + 15 + 6x + 15}{2x^2 + 7x + 5}$ $\dfrac{6x^2 + 27x + 30}{2x^2 + 7x + 5}$ then factorise to get $\dfrac{3(2x + 5)(x + 2)}{(2x + 5)(x + 1)}$ cancel the $(2x + 5)$	$\dfrac{3(x + 2)}{x + 1}$

	Reflection in the x-axis	Reflection in the y-axis	Translation
Reflections and Translations of functions	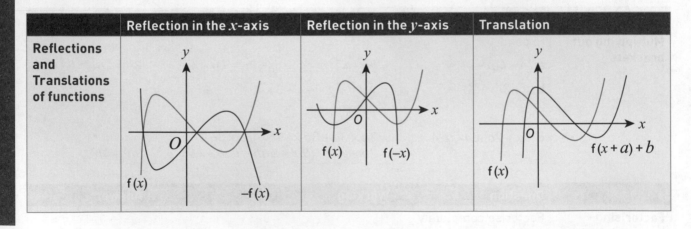		

Geometry and Measures

Rules

The **perimeter** of a shape is the distance around its edge. You **add** all the side lengths.

The **area** of a shape is the amount of flat surface it has. You **multiply** two lengths.

The **volume** of a shape is the amount of space it has. You **multiply** three lengths.

Alternate angles are in the shape of a letter **Z**.

Corresponding angles are in the shape of a letter **F**.

Allied angles or co-interior angles are in the shape of a letter **C**.

	Question	Working	Answer
Perimeter of a shape	Find the circumference of a circle diameter 5 cm.	$C = \pi D$ $C = \pi \times 5$ You have used one length.	5π cm or 15.7 cm

	Question	Working	Answer
Area of a shape	Find the area of a circle with radius 5 cm.	$A = \pi r^2$ $A = \pi \times 5^2$ or $\pi \times 5 \times 5$ You have multiplied 2 lengths.	25π cm² or 78.5 cm²

	Question	Working	Answer
Volume of a solid	Find the volume of this shape with radius 5 cm and height 12 cm.	For this cylinder you need to use the formula Volume = $\pi \times r^2 \times h$ So volume is $\pi \times 5 \times 5 \times 12 = 300\pi$ You have multiplied 3 lengths.	300π cm³

	Question	Working & Answer
Angles between parallel lines	Find the missing angles in this diagram. Give reasons for your answer. 50° a b c	$a = 50°$ (Alternate angles are equal) $b = 130°$ (Allied angles add to 180° (supplementary)) $c = 50°$ (Corresponding angles are equal)

	Question	Working & Answer
Finding missing angles and giving reasons	TAP is a tangent to the circle. ABC is an isosceles triangle. Find angle $x°$.	Angle ACB = 50° (Angle between a tangent and a chord is equal to the angle in the alternate segment) Angle ABC = (180 − 50) ÷ 2 = 65° (Angles in a triangle add to 180° and Base angles of an isosceles triangle are equal) Angle x = 180 − 65 = 115° (Opposite angles of a cyclic quadrilateral add to 180° (supplementary))

	Question	Working & Answer
Similar shapes	David makes similar statues with their heights in the ratio 4:5. **a** The surface area of the small statue is 100 cm². Find the surface of the large statue. **b** The volume of the large statue is 75 cm³. Find the volume of the small statue.	**a** Linear scale factor is 4:5 Area scale factor is $4^2:5^2 = 16:25$ Area of small statue is 100 cm² Area of large statue is $100 \times \frac{25}{16} = 156.25$ cm² **b** Linear scale factor is 4:5 Volume scale factor is $4^3:5^3 = 64:125$ Volume of large statue is 75 cm³ Volume of small statue is $75 \times \frac{64}{125} = 38.4$ cm³

Statistics and Probability

	Question	Working	Answer
Mean from a grouped frequency table	Work out an estimate of the mean age from this frequency table 	Age	f	 | $0 < a \leqslant 10$	4	 | $10 < a \leqslant 20$	6	 | $20 < a \leqslant 30$	12	 | $30 < a \leqslant 40$	5	 | $40 < a \leqslant 50$	3		Multiply the mid value of the age groups by the frequency $5 \times 4 = 20$ $15 \times 6 = 90$ $25 \times 12 = 300$ $35 \times 5 = 175$ $45 \times 3 = 135$ Divide the **total** of age × frequency by the **total frequency** 720 ÷ 30 Note: Don't forget to divide by the total frequency (30) and not the number of groups (5).	24

	Question	Working	Answer
Pie charts	Draw a pie chart from this information 	Favourite colour	f
---	---
Red	7
Blue	4
Green	2
Yellow	3
Black	4		As pie charts are based on a circle then we need to divide the number of degrees in a whole turn (360°) by the total frequency which is 20. So 360° ÷ 20 = 18° The angle for each colour is then calculated by multiplying its frequency by 18°.	Red $\quad 7 \times 18° = 126°$ Blue $\quad 4 \times 18° = 72°$ Green $2 \times 18° = 36°$ Yellow $3 \times 18° = 54°$ Black $\quad 4 \times 18° = 72°$ Then draw the circular pie chart.

	Question	Working	Answer
Histograms	Draw a histogram from this information 	lengths	f
---	---
$0 < h \leqslant 30$	9
$30 < h \leqslant 50$	12
$50 < h \leqslant 60$	20
$60 < h \leqslant 80$	18
$80 < h \leqslant 100$	10		Histograms represent frequencies by area. You divide the frequency by the group width to get the height of the frequency density e.g. 9 ÷ 30 = 0.3, 12 ÷ 20 = 0.6 etc.	Column heights of bars are | lengths	h
---	---
$0 < h \leqslant 30$	0.3
$30 < h \leqslant 50$	0.6
$50 < h \leqslant 60$	2.0
$60 < h \leqslant 80$	0.9
$80 < h \leqslant 100$	0.5	

Common areas where students make mistakes

One week to go

You need to know these formulas and essential techniques.

Number

Topic	Formula	When to use it
Order of operations	BIDMAS	If you have to carry out a calculation. You use the order Brackets, Indices, Division, Multiplication, Addition and Subtraction.
Simple interest	SI for 5 years at 3% on £150 $$\frac{3}{100} \times 150 \times 5$$	To find the **Simple interest** you find the interest for one year and multiply by the number of years.
Compound interest	CI for 2 years at 3% on £150 Year 1 $\frac{3}{100} \times 150 = £4.50$ Year 2 $\frac{3}{100} \times (150 + 4.50)$	For **Compound interest** you find the percentage interest for one year, add it to the initial amount and find the interest on the total and so on. You can also do this using geometric progressions and write £150 $\times$ (1.03)^2 =
Standard form	$2.5 \times 10^3 = 2500$ $2.5 \times 10^{-3} = 0.0025$	A number in standard form is (a number between 1 and 10) × (a power of 10)
Approximating	Decimal places	You round to a number of decimal places by looking at the next decimal place and rounding up or down.
	Significant figures	The first non-zero digit is always the first significant figure and you count the number of significant figures then look at the next figure and round up or down. You should always keep the idea of the size of the number.
Recurring decimals	Change $0.\dot{1}\dot{8}$ into a fraction	Let $x = 0.\dot{1}\dot{8}$ then $100x = 18.\dot{1}\dot{8}$ (Multiply by 10 to the power of the number of recurring digits); 2 this time Subtracting gives $99x = 18$ so fraction is $\frac{18}{99}$ or $\frac{2}{9}$

Algebra

Topic	Formula	When to use it
Rules of indices	$n^a \times n^b = n^{a+b}$	When you multiply you add the indices or powers.
	$n^a \div n^b = n^{a-b}$ $\left(n^a\right)^b = n^{a \times b}$	When you divide you subtract the indices or powers.
	$n^{\frac{1}{2}} = \sqrt{n}$ $\quad n^{\frac{1}{3}} = \sqrt[3]{n}$	When you raise a power to a power you multiply the indices or powers.
	$n^{-a} = \frac{1}{n^a}$	A fractional index is a root. A negative index means the reciprocal.
	$\sqrt{a \times b} = \sqrt{a} \times \sqrt{b}$	The square root of a product is the product of the square roots.
	$\sqrt{\frac{a}{b}} = \frac{\sqrt{a}}{\sqrt{b}}$	The square root of a division is the division of the roots.
	$(a-b)(a+b) = a^2 - b^2$	Difference of 2 squares often appears.

Straight line graph	$y = mx + c$	m is the gradient and $(0, c)$ the intercept on the y-axis.
Parallel and perpendicular lines	$y = mx + c$ and $y = mx + d$	are parallel lines because the gradients are equal.
	$y = mx + c$ and $y = -\frac{1}{m}x + d$	are perpendicular lines because the gradients multiply to –1.
Curved line graphs	$y = ax^2 + bx + c$ where $a \neq 0$	Is a quadratic graph. When a is +ve shape is U. When a is –ve shape is ∩.
	$y = ax^3 + bx^2 + cx + d$ where $a \neq 0$	Is a cubic graph and is in the shape of an ∿.
	$y = \frac{k}{x}$	Is a reciprocal graph. It is in 2 parts and the curved lines approach asymptotes.
	$y = ab^x$	Is a growth function when x is positive and decaying function when x is negative.
	$x^2 + y^2 = r^2$	Is the equation of a circle centre the origin and radius r.
Quadratic equations	$x = \frac{-b \pm \sqrt{b^2 - 4ac}}{2a}$	This quadratic equation formula is used to solve a quadratic equation of the form $ax^2 + bx + c = 0$. You just substitute the values of a, b and c into the formula.
Linear series	$a + (n - 1)\,d$ is the nth term	a is the first term and d the difference between each term.
Graph transformations	When f(x) becomes –f(x)	It is a reflection in the x-axis.
	When f(x) becomes f($-x$)	It is a reflection in the y-axis.
	When f(x) becomes f($x - a$) + b	It is a translation of $\begin{pmatrix} a \\ b \end{pmatrix}$.

Geometry and Measures

Topic	Formula	When to use it
Parallel sides	→	Parallel lines are shown with arrows.
Equal sides		Equal lines are shown with short lines.
Perimeter	Add lengths of all sides.	To find the perimeter of any 2D shape.
Areas of 2D shapes	Area = $l \times w$	Area of a rectangle is length × width
	Area = $\frac{1}{2}b \times h$	Area of a triangle is $\frac{1}{2}$ base × vertical height
	Area = $b \times h$	Area of a parallelogram is base × vertical height
	Area = $\frac{1}{2}(a + b) \times h$	Area of a trapezium is $\frac{1}{2}$ the sum of the parallel sides × the vertical height

Circumference and area of a circle	$C = \pi \times D$ or $C = \pi \times 2r$ $A = \pi \times r^2$	Circumference or the perimeter of a circle is: pi × diameter or pi × double the radius. Area of a circle is: pi × radius squared	
Volumes of 3D shapes	$V = l \times w \times h$ $V = \pi r^2 h$	Volume of a cuboid is: length × width × height Volume of a cylinder is: area of circular end × height	
Volumes of prisms	Volume of a prism = area of end × length	Multiply the area of cross section by the length.	
Volumes of pyramids	Volume of a pyramid: $\frac{1}{3}$ area of base × height	Multiply the area of the base by the vertical height and divide by 3	
Volume of a cone	Volume of a cone is: $V = \frac{1}{3}\pi r^2 h$	one third area of circular base × height	
Pythagoras' theorem	$c = \sqrt{a^2 + b^2}$	The hypotenuse of a right-angled triangle can be found by finding the square root of the sum of the squares of the two shorter sides. One of the shorter sides of a right-angled triangle can be found by finding the square root of the difference between the hypotenuse squared and the other shorter side squared.	
Trigonometry, right-angled triangles	$\sin = \frac{o}{h}$; $\cos = \frac{a}{h}$; $\tan = \frac{o}{a}$	You can find a missing side or a missing angle by selecting and using one of these formulae. You use the trigonometry ratio that has two given pieces of information and the one you have to find.	
Trigonometry, with triangles that are not right-angled	$c^2 = a^2 + b^2 - 2ab\cos C$ $\dfrac{a}{\sin A} = \dfrac{b}{\sin B} = \dfrac{c}{\sin C}$ Area $= \frac{1}{2} ab \sin C$	You use the Cosine rule when you have a triangle with 2 sides and the angle between them. You use the Sine rule when you have 2 sides or 2 angles with one to find. You use the area of a triangle formula when you have 2 sides and the angle between them. Use the cosine rule to find a side given 2 sides and an included angle or an angle if given 3 sides.	

Statistics and Probability

Topic	Formula	When to use it
Probability	$P(A \text{ and } B) = P(A) \times P(B)$	You use this when you have two independent events
	$P(A \text{ or } B) = P(A) + P(B)$	You use this when you have mutually exclusive events
	$P(A \text{ or } B) = P(A) + P(B) - P(A) \times P(B)$	You use this when you do not have mutually exclusive events

Answers

Number

Number: pre-revision check (page 1)

1 a $5357.142\ldots$ or $5.357\ldots \times 10^3$

b 2302 or 2.302×10^3

2 a $1.03\dot{7}$ **b** $\frac{2}{11}$

3 lower bound = 8.365, upper bound = 8.375

4 a 3.22, 3.24   **b** 13.391775, 13.471875

5 £112 000   **6** £8603   **7** 6 years

8 A = inverse, B = direct, C = inverse

9 $T = \frac{35}{x}$ **10** $P = 6\sqrt{A}$ **11 a** 4   **b** 10 000

12 a 9   **b** 0.2   **13 a** $10\sqrt{7}$ **b** $\frac{\sqrt{3}}{6}$

Calculating with standard form (page 2)

1 a 4.5188×10^3 **b** $3.994\ldots \times 10^9$

2 0.29 nanometres   **3** $6.324\ldots \times 10^4$

Recurring decimals (page 3)

1 $100x - 10x = 54.4444\ldots - 5.4444\ldots$; $90x = 49$

2 $1000x - x = 425.425\,425\ldots - 0.425\,425\ldots$; $999x = 425$

3 $2.\dot{1}\dot{8} - 1.\dot{1} = 1.0\dot{7}$ followed by standard proof

Rounding to decimal places, significance and approximating (page 4)

1 11.44 cm²   **2** 11.5 cm   **3** 2000

Limits of accuracy (page 5)

1 a 2.25 m and 1.15 m   **b** 2.35 m and 1.25 m

2 2495 m or 2.495 km

3 No. The average speed is between $61\frac{2}{3}$ and 65 mph

Calculating with lower and upper bounds (page 6)

1 $\text{LB} = \sqrt{\frac{53}{3.145}} = 4.1051\ldots$ $\text{UB} = \sqrt{\frac{55}{3.135}} = 4.1885\ldots$

Length radius = 4 cm

2 $62.25 - 58.75 = 3.5$ s

Reverse percentages (page 7)

1 £320   **2** No, he had 5019 in 2014

3 Better off by £234.89

Repeated percentage increase/decrease (page 8)

1 £4589.96

2 No, since the car's value will be £8109.52 after 5 years while half the cost is £8750

3 3 years

Growth and decay (page 9)

1 15.1 °C   **2** 4 years

Mixed exam-style questions (page 9)

1 Noreen is correct.

2 a 30.2   **b** 7.81×10^8 miles

3 $100x - 10x = 572.222\ldots - 57.222\ldots = 515$; $90x = 515$; $x = 5\frac{15}{90} = 5\frac{13}{18}$

4 Yes, since $30.5 \times 18.5 = 564.25$ cm²

5 Yes, since $3.55 \div \left(\frac{7.25}{60}\right) = 29.38\ldots$ i.e. LB ÷ UB is less than 30   **6** £840

7 The increase is not 500 each year; the percentage increase is 3.2258% p.a., giving 21 2912 population after 10 years.   **8** 2930 fish

Working with proportional quantities (page 11)

1 £9.75   **2** £1.35   **3** 88

The constant of proportionality (page 12)

1 a 4.00, 8.10, 8.00   **b** $E = 1.35P$

c the exchange rate   **2** 87.5

Working with inversely proportional quantities (page 13)

1 a inverse, since $8 \times 25 = 10 \times 20 = 200$

b 50 days

2 $x = 25, y = 12$ (not 120)   **3** 10

Formulating equations to solve proportion problems (page 14)

1 a $D = 5t^2$ **b** 4 seconds

2

x	1	8	64	216	1000
y	2400	1200	600	400	240

3 1600 units

Index notation and rules of indices (page 15)

1 a 10^4 **b** 32 768   **2** 2^{15}

3 $10^4 \times 10^5 = 10^9$; $10 \times 10^2 = 10^3$; $\frac{10^{20}}{10^2} = 10^{18}$; $10^{10} = 10\,000\,000\,000$

Fractional indices (page 16)

1 a 8   **b** 0.25

2 $25^{-\frac{1}{2}}$ $16^{-\frac{1}{4}}$ 27^0 $\frac{1}{4^{-\frac{1}{2}}}$ $81^{\frac{3}{4}}$ **3** $\frac{11}{6}$

Surds (page 17)

1 a $\frac{\sqrt{21}}{7}$ **b** $2\sqrt{5} + 4$

2 $\frac{32\sqrt{3}}{27}$ **3** $7 + 5\sqrt{2}$

Mixed exam-style questions (page 18)

1 a $y = kx$ and $x = cz$, so $y = kcz = \text{constant} \times z$

b 50

2 a t is inversely proportional to s.   **b** $3\frac{1}{3}$

3 281.25 m   **4** −2.5   **5** 1254

6 a 0.1   **b** 0.4   **7** $2\sqrt{5}$

Algebra

Algebra: pre-revision check (page 19)

1 a i a^{10} **ii** x^3 **iii** $\dfrac{3f^2}{2e^3}$

   b i $t^2 + 7t + 10$ **ii** $v^2 - 2v - 35$ **iii** $y^2 - 11y + 30$

2 a 100 **b** $a = \dfrac{2(s - ut)}{t^2}$

3 $n - 1 + n + n + 1 = 3n$. This is a multiple of 3 since 3 is always a factor

4 a $\dfrac{a^3 b^4}{c^2}$ **b** $\dfrac{a^{\frac{3}{2}} c^{\frac{1}{4}}}{b^{\frac{3}{2}}}$ **5** $x = 1.25$

6 a 34.6 to 3s.f.   **b** $x = \dfrac{3y - 5}{1 + 2y}$ **7** 30

8 a 3 **b** $98\,415$

9 $2n^2 - 3n + 4$ **10** $y = -\dfrac{1}{2}x + \dfrac{3}{2}$

11 a Sketch drawn through $(0, 3)$, $(1, 0)$, $(3, 0)$, minimum at $(2, -1)$

   b $x = 1$ and $x = 3$ **c** $x = 2$

12 a Curve drawn through $(-3, -9)$, $(-2, 2)$, $(-1, 3)$, $(0, 0)$, $(1, -1)$, $(2, 6)$, $(3, 27)$

   b $x = -2.3$ or $x = 0$ or $x = 1.3$

13 $y = -\dfrac{1}{2}x + 5$ **14 a** $-3 \leqslant x < 4$

   b i $x < 2$ **ii** $t > 1.2$ **iii** $y \leqslant 4.5$

15 $x = 2$, $y = -1$ **16** $x = \dfrac{1}{2}$, $y = 2$

17 Lines $x + y = -1$, $y = 1 - 2x$ and $y = x + 3$ accurately plotted, and the triangular region between the lines shaded.

18 a i $x^2 - x - 20$ **ii** $y^2 - 64$ **iii** $36 - a^2$

   b i $(x + 3)(x + 4)$ **ii** $(e + 2)(e - 5)$

   iii $(b + 5)(b - 5)$

19 a $x = 2$ or $x = 3$ **b** $x = -3$ or $x = 5$

   c $p = -7$ or $p = 7$

20 a i $(2x - 1)(2x + 3)$ **ii** $(3b + 8)(3b - 8)$

   b i $x = -5$ or $x = \dfrac{4}{3}$

     ii $x = -\dfrac{3}{2}$ **c** $\dfrac{1}{x - 5}$

21 a $x = -1.366$ or $x = 0.366$

   b $x = -10.424$ or $x = -0.576$

22 a About $4.4\,\text{m/s}^2$ **b** $3.75\,\text{m/s}^2$

23 a Accurate drawing.   **b** Accurate drawing.

24 About $326\,\text{m}$

Simplifying harder expressions and expanding two brackets (page 21)

1 a $\dfrac{3y^6}{4}$ **b** $\dfrac{5a^3 b^2}{4}$

2 $(2a + 5)(a + 3) - a^2 = 2a^2 + 6a + 5a + 15 - a^2$
$= a^2 + 11a + 15$

Using complex formulae and changing the subject of a formula (page 22)

1 -540 **2** $t = \sqrt{\dfrac{y - 3s}{5a}}$

Identities (page 23)

1 $x^2 - 7x + 12 = (x - 3)(x - 4)$
so $p = -3$ and $q = -4$ or vice versa

Using indices in Algebra (page 24)

1 a $p^2 q^{\frac{3}{2}} r^{\frac{1}{2}}$ or $p^2 \sqrt{q^3 r}$ **b** $x^{\frac{11}{6}} y^{-\frac{7}{4}}$ or $\dfrac{\sqrt[6]{x^{11}}}{\sqrt[4]{y^7}}$

2 $n = -\dfrac{16}{3}$

Manipulating more expressions; algebraic fractions and equations (page 25)

1 a $\dfrac{5x^2 + 23x - 12}{x^2 - 16}$ **b** $x = \dfrac{25}{13}$

2 $(n + 1)^3 - (n + 1)^2$
$= (n + 1)(n^2 + 2n + 1) - (n^2 + 2n + 1)$
$= n^3 + 3n^2 + 3n + 1 - (n^2 + 2n + 1)$
$= n^3 + 3n^2 + 3n + 1 - n^2 - 2n - 1$
$= n^3 + 2n^2 + n$
$= n(n^2 + 2n + 1)$
$= n(n + 1)^2$

Rearranging more formulae (page 26)

1 $m = \dfrac{P - 8c}{2 - 3c}$ or $\dfrac{8c - P}{3c - 2}$ **2** $T = \dfrac{K^2 S}{P - K^2}$

Special sequences (page 27)

1 a $n^2 + 1$ **b** 401 **2** $(n + 1)(n + 2)$

Quadratic sequences (page 28)

1 14 and 34 **2** $2n^2 + 5$

*n*th term of a quadratic sequence (page 29)

1 $3n^2 - 2n + 1$ **2** $n = 1$ so number is 5

The equation of a straight line (page 30)

1 $y = 2x - 1$

2 P, S and T are parallel to each other and so are Q and R

Plotting quadratic and cubic graphs (page 31)

1 a by inspection   **b** $x = 1$ or $x = 3$

2 by inspection

Finding equations of straight lines (page 32)

1 $y = 3x + 3$ **2** $y = -2x + 1$

Polynomial and reciprocal functions (page 33)

1 a Accurate graph.

   b All three meet at $(0,0)$, $(1, 1)$ and $y = x$ and $y = x^3$ also meet at $(-1, -1)$

2 Accurate graph. Fuel consumption approaches 60 as speed increases

Perpendicular lines (page 34)

1 $4y + x = 6$ or $y = -\dfrac{1}{4}x + \dfrac{3}{2}$

2 5 square units

Exponential functions (page 35)

1 a The difference between each year is: 330, 262, 208, 165, 131, 104. This means the rate of decrease is decreasing each month. Every 3 years the population halves so the population is decreasing exponentially.

b $P = 1600 \times 2^{-\frac{t}{3}}$ or $P = 1600 \times \left(\frac{1}{2}\right)^{\frac{t}{3}}$

Trigonometric functions (page 37)

1 20°, 100°, 140°, 220°, 260°, 340°

2 a $t = 6$ and $t = 18$ **b** $t = 0$, $t = 12$ and $t = 24$
c $t = 3, 9, 15, 21$

Mixed exam-style questions (page 38)

1 10:20 a.m.

2 $4(10x + 15) = 5(8x + 12)$;
$10x + 15 - (8x + 12) = 2x + 3$

3 Area of large square $= 36x^2 - 12x - 1$

Area of orange square is $36x^2 - 12x + 1 - (10x^2 + 14x - 12) = 26x^2 - 26x + 13 = 13(2x^2 - 2x + 1)$

4 a $n + n + 1 + n + 2 + 10 + n + 1 + 20 + n + 1$
$= 5n + 35 = 5(n + 7)$

b If $5n + 35 = 130$; then $n = 19$. N cannot equal 19 because it will overlap the grid.

5 $2n^2 - 3n + 2$ **6 a** $x = \dfrac{20\pi^2}{y^2 - 12\pi^2}$

b LCM is $72a^3b^3c^4$. HCF is $6a^2b^2c$.

7 $\dfrac{x + 4}{2x + 5}$ **8** $n = -\dfrac{1}{3}$ **9 a** $\dfrac{1}{32}$ **b** $n \geq 12$

10 a £15

b Draw line on graph from $(0, 0)$ to $(5, 125)$ or compare cost for each day. Up to 2 days cheaper with **Car Co**, 3 days or more is cheaper with **Cars 4 U**, for 2 days both companies charge the same (£50).

11 a $y = 2x - 7$

b Substituting in the values of x and y from $(3, -1)$ gives $y = -1$ and $2 \times 3 - 7 = 6 - 7$ which also gives -1. Therefore the point $(3, -1)$ lies on the line l. **c** $y = -\dfrac{1}{2}x - 2$

12 a $y = -1$ **b** $x = 3.3, y = 0.7$ and $x = -0.3, y = 4.3$

13 a Accurate graph.
b $y = (x - 2)^2$ or $y = x^2 - 4x + 4$

14 Cathy is 24.

15 Graph drawn and co-ordinates found are $(4\frac{1}{3}, \frac{2}{3})$, $(\frac{2}{3}, 4\frac{1}{3})$, and $(-3, -3)$

16 $(x - 8)$ is the width and $(x - 4)$ is the length so perimeter is $4x - 24$

17 a $4y = 3x - 8$ **b** $4y = 3x + 5$ **c** $3y + 4x + 1 = 0$

18 Between 06:00 and 11:00

Trial and improvement (page 40)

1 2.51

Linear inequalities (page 41)

1 a

b $y \leqslant 3$ **2** A number less than 10

Solving simultaneous equations by elimination and substitution (page 42)

1 $a = 2$ and $b = 3$

2 12 ordinary coaches and 3 superior coaches

Using graphs to solve simultaneous equations (page 43)

1 $x = -1$ and $y = 1$

2 Peach is cheaper up to 40 Mbytes.
At 40 Mbytes both companies charge £40.
After 40 Mbytes M−mobile is cheaper.

Solving linear inequalities (page 45)

1 Accurate drawing.

Factorising quadratics of the form $x^2 + bx + c$ (page 46)

1 $x^2 + 6x + 8 = (x + 2)(x + 4)$
2 $x^2 - 2x - 8 = (x + 2)(x - 4)$
3 $x^2 + 2x - 8 = (x - 2)(x + 4)$
4 $x^2 - 6x + 8 = (x - 2)(x - 4)$
5 $x^2 - 16 = (x + 4)(x - 4)$

Solve equations by factorising (page 47)

1 a $x = -6$ or $x = 2$ **b** $x = 0$ or $x = 5$
c $x = 9$ or $x = -2$ **d** $x = +5$ or $x = -5$

2 Ben could use 2 or −12

Factorising harder quadratics and simplifying algebraic fractions (page 48)

1 $x = 2\dfrac{1}{3}$; shortest side = 3 cm **2** $\dfrac{4}{3x + 4}$

The quadratic equation formula (page 49)

1 $x = 1.74$ or -0.34

2 The equation becomes $2x^2 - 8x + 8 = 0$
The value of $b^2 - 4ac$ is $(-8)^2 - 4 \times 2 \times 8$
$= 64 - 64 = 0$
Since this is 0 there is only one solution or two solutions which are the same

Using chords and tangents (page 51)

1 a Draw the graph and find the gradient when $t = 5$. Answer is 5 m/s

b Find the gradient of the line joining $(0, 0)$ to $(5, 12.5)$. Answer is 2.5 m/s

Translations and reflections of functions (page 52)

1 Accurate graphs.

2 If f(x) is reflected in the y-axis then it becomes f(−x). So $y = x^3 - 4x$ becomes $y = (-x)^3 - 4(-x)$ or $y = -x^3 + 4x$ or $y = 4x - x^3$.
When this is translated by $\begin{pmatrix} 2 \\ -3 \end{pmatrix}$ since f(x) becomes f(x − 2) − 3.
Which makes: $y = 4(x - 2) - (x - 2)^3 - 3$
$y = 4x - 8 - (x^3 - 6x^2 + 12x - 8) - 3$
$y = 4x - 8 - x^3 + 6x^2 - 12x + 8 - 3$
$y = -x^3 + 6x^2 - 8x - 3$

Area under non-linear graphs (page 54)

1 Accurate graph. Area $= \frac{1}{2} \times 1.8 \times 1 + \frac{1}{2}(1.8 + 3.2) \times 1 + \frac{1}{2}(3.2 + 4.2) \times 1 + \frac{1}{2}(4.2 + 4.8) \times 1 + \frac{1}{2}(4.8 + 5) \times 1 = 0.9 + 2.5 + 3.7 + 4.5 + 4.9 = 16.5$ metres

Mixed exam-style questions (page 55)

1 $x = 5.27$ or $x = -1.27$

2 When $x = 2$, $x^3 - 3x - 5 = 2^3 - 3 \times 2 - 5 = -3$
When $x = 3$, $x^3 - 3x - 5 = 3^3 - 3 \times 3 - 5 = +13$
so there is a root between $x = 2$ and 3
$x^3 - 3x - 5 = 0$ so $x^3 = 3x + 5$ and $x_2 = \sqrt[3]{3x + 5}$
When $x_1 = 2$, $x_2 = \sqrt[3]{3x + 5} = \sqrt[3]{11} = 2.223980091$
When $x_2 = 2.223980091$, $x_3 = 2.268372388$
When $x_3 = 2.268372388$, $x_4 = 2.276967162$
When $x_4 = 2.276967162$, $x_5 = 2.278623713$
When $x_5 = 2.278623713$, $x_6 = 2.278942719$
So $x = 2.279$ to 3 decimal places

3 a $\frac{5}{n-2} - \frac{2}{n+2} = \frac{5(n+2) - 2(n-2)}{(n-2)(n+2)}$

$= \frac{5n + 10 - 2n + 4}{n^2 - 4} = \frac{3n + 14}{n^2 - 4}$

b There can be no answer to this final expression if you divide by zero. If $n^2 - 4 = 0$ then $n^2 = 4$ so $n = \pm 2$ are the two values.

4 29.3 cm **5 a** $\frac{25}{40} = 0.625 \text{ ms}^{-2}$

b $-\frac{15}{40} \text{ ms}^{-2} = -0.375 \text{ ms}^{-2}$ **c** 1680 m
from $10(0 + 0 + 2(20 + 25 + 25 + 10 + 4))$

6 (5, 2) and (1.4, −5.2)

7 a The difference between each year is: 2, 3, 4, 6, 8, 12. This means the rate of increase is increasing each month. Every 2 years the population doubles so the population is increasing exponentially **b** $P = 5 \times 2^{\frac{t}{2}}$

Geometry and Measures

Geometry and Measures: pre-revision check (page 57)

1 14.98 g **2** A and C, SAS

3 All equilateral triangles have angles of 60°, therefore all equilateral triangles are similar. Equilateral triangles can have different length sides, therefore not all equilateral triangles are congruent.

4 a $a = 90°$, the angle in a semicircle is 90°.
b $b = 40°$, the angle between a chord and a tangent is equal to the angle in the alternate segment.

5 8.1 cm (1 d.p.) **6 a** 14.7 cm (3 s.f.)
b 51.3 cm² (3 s.f.)

7 14.2 cm (3 s.f.) **8** 17.9 cm (3 s.f.)

9 a circle radius 3 cm drawn around a point A
b line equidistant between parallel lines

10 $x = 6$ **11** 6.1 cm to 1 d.p. **12** $7\sqrt{3}$ cm

13 90° anti-clockwise rotation about (3, −1)

14 enlargement scale factor −2 , centre of enlargement (1, 2) **15** 48.6°

16 Accurate drawing of plan, front and side.

17 a 254 mm³ (3 s.f.) and 226 mm² (3 s.f.)
b 163 cm³ (3 s.f.)

18 52 cm²

Working with compound units and dimensions of formulae (page 59)

1 3.72 person/km² increase
2 a 7.78 g/cm³ **b** 11 667 kg or 11 700 to 3 s.f.
3 That it's travelling at 64.6 or 65 miles/hr

Congruent triangles and proof (page 60)

1 AC is common to AYC and AXC. AX = CY, given. AY = CX, perpendicular bisector of an equilateral triangle. So AYC and AXC are congruent (SSS).

2 SQ common to XQS and QYS. PQ = SR (opposite sides of a parallelogram are equal) therefore XQ = SY as X and Y are mid points of PQ and RS. Angle XQS = angle QSY, alternate angles. So XQS and QYS are congruent (SAS).

Proof using similar and congruent triangles (page 61)

1 AXD = BXC, vertically opposite angles
ADX = XBC, alternate angles
DAX = XCB, alternate angles
Both triangles have three equal angles so they are similar

2 PT = TS = QR = RS, sides of a regular pentagon are equal. Angles PTS and QRS are equal, interior angles of a regular pentagon are equal. Therefore triangles PTS and QRS are congruent (SAS). As triangles PTS and QRS are congruent, PS = QS, therefore PQS is isosceles as it has 2 equal sides.

Circle theorems (page 62)

1 a $a = 90°$, angle in a semicircle = 90°;
$b = 53°$, angles in a triangle;
$c = 37°$, angle between a chord and a tangent is equal to the angle in the alternate segment
b $d = 63°$ angle at circumference is twice angle at centre;
$e = 117°$, opposite of a cyclic quadrilateral add up to 180°

2 Angle RQT = angle RQP (RQ bisects PQT, given)
Angle RPQ = angle RQT, (angle between a chord and a tangent is equal to the angle in the alternate segment). Angle RQP = angle RPQ, so triangle PQR is isosceles and RP = RQ.

Pythagoras' theorem (page 63)

1 7.6 cm **2** 3855 m

Arcs and sectors (page 64)

1 31 cm **2** 2.57 cm²

The cosine rule (page 66)

1 35° **2** 386 km

The sine rule (page 67)

1 096° **2** 93.53 cm²

3 a 4.3 cm **b** 43.0 cm²

Loci (page 68)

1 Circle radius 5 cm at A, circle 3.5 cm radius at B. Accurate drawing.

2 Intersection of perpendicular bisectors of 3 sides of XYZ

Mixed exam-style questions (page 69)

1 AB = AC, BD = EC (given)
Angle ABC = angle ACB (isosceles triangle), so ABD is congruent with ACE: therefore AD = AE so triangle ADE is isosceles

2 Angle QPX = angle XTS (alternate angles), angle STX = angle XPQ (alternate angles), angle PXQ = angle SXT (vertically opp). PQX is similar to SXT (AAA) Corresponding sides equal, therefore triangles are congruent

3 a 2.12 cm (3 s.f.) **b** 1.28 cm² (3 s.f.)

4 a 10.5 cm (3 s.f.) **b** 3.33 cm (3 s.f.)

5 Angle QPX = angle XRS (alternate angles), angle RSX = angle XQP (alternate angles), angle PXQ = angle SXR (vertically opp). PQX is similar to SXR (AAA)

6 9.49 cm

7 a ABX = ACD, angles in same segment
BAC = BDC, angles in same segment
AXB = DXC, vertically opp, therefore ABX and DXC are similar (AAA)

b AT = TD tangents at a point
Triangle TAD is isosceles
ADT = DAT = 50°
PAB = ADB = 20°, angles in alternate segment are equal
ADT = ABD = 50°, angles in alternate segment are equal
ACD = ABD = 50°, angles on same arc are equal
BCA = ADB = 20°, angles on same arc are equal
BCD = 50 + 20 = 70°
BAD = 110°, angles on a straight line add up to 180°
BAD + BCD = 180°, ABCD must be cyclic as opposite angles are supplementary.

8 255° nearest degree

9 a 40.3 cm³ (1 d.p.) **b** 89.3 cm² (1 d.p.)

Similarity (page 71)

1 4 cm **2** 13.5 cm

Trigonometry (page 72)

1 a 21.8° **b** 3.2 m

2 12.4 cm

Finding centres of rotation (page 74)

1 a (−2, 0)

b 90° anti-clockwise rotation about (−2, 0)

2 90° clockwise rotation about (−1, −2)

Enlargement with negative scale factors (page 76)

1 enlargement with scale factor −2, centre of enlargement (3, −1)

2 a and b Accurate drawing.

c enlargement scale factor $\frac{2}{3}$, centre of enlargement (1, −1)

3 rotation 180°, centre of rotation (x, y)

Trigonometry in 2D and 3D (page 77)

1 70.5° (1 d.p.) **2 a** 35.3° **b** 54.7°

3 2.8 cm (1 d.p.)

Volume and surface area of cuboids and prisms (page 78)

1 a 98.2 m³ **b** 9817 litres

2 a 6 cm² **b** 27 cm³ **c** 66 cm²

3 24 cm × 2 cm × 2 cm. 4 cm × 4 cm × 6 cm. 4 cm × 12 cm × 2 cm. 2 cm × 8 cm × 6 cm

Enlargement in two and three dimensions (page 80)

1 a 810 cm² **b** 1 : 27 **2** 54 cm³ **3** 4.9 cm

Constructing plans and elevations (page 81)

1 Accurate drawing.

2 a Accurate drawing of plan, front and side.

b Accurate drawing of plan, front and side.

Surface area and 3D shapes (page 82)

1 6 cm

2 a 15 000 cm³ to 3 s.f. **b** 3900 cm² to 3 s.f.

Area and volume in similar shapes (page 83)

1 1584 cm² **2** 19.3 cm **3** 212 cm²

Mixed exam-style questions (page 84)

1 5.9 km **2** 2.09 g

3 approx ratio Moon to Earth 1 : 13.45

4 a 35.1 cm² **b** 526 cm³ **c** 3.90 cm²

d 52.8 cm² (all 3 s.f.) **5** 3.22 m

6 a accurate scale drawing of net

b answers using measured height in range 50 cm²–70 cm²

7 32.67 minutes

8 a Accurate drawing.

b Area = 18(front) + 24(back) + 21(sides) + $6\sqrt{10}$ (roof) = 81.97 m² 6.83 litres, therefore 2 tins required.

9 rotation 180° about (0, 1)

10 a 302 mm² **b** 148 mm² **c** 14 g

Statistics and Probability: pre-revision check (page 86)

1 a 0.9

b Median LL > median CL, i.e. on average the weight of the fish in LL is greater than the weight of the fish in CL.
IQR LL = IQR CL

2 a Accurate scatter diagram.

b Positive correlation. The older the tree the greater the trunk radius

c i 50 – 65 cm   **ii** Not reliable – extrapolation

3 Accurate histogram.   **4** $\frac{14}{30}$

5 a Accurate Venn diagram.   **b** $\frac{3}{50}$ **6** $\frac{1}{8}$

Using grouped frequency tables (page 88)

1 61.29 seconds (2 d.p.)

2 a $1 < w \leqslant 1.5$ **b** $1.5 < w \leqslant 2$

   c 1.73 kg (2 d.p.)

Inter-quartile range (page 90)

1 a Accurate cumulative frequency diagram.

b 1.5 °C

2 On average the times taken in 2015 were greater than the times taken in 1995 (as the median time taken in 2015 > the median time taken in 1995). The times taken in 1995 were more consistent than the times taken in 2015 (as inter-quartile range for 2015 > inter-quartile range for 1995). The distribution of times in 1995 are symmetrical, but in 2015 they are negatively skewed.

Displaying grouped data (page 92)

a Time taken can take any value in a given interval.

b Accurate frequency table.

c Accurate frequency diagram.

d The median is the $\frac{25+1}{2}$ = 13th value in the ordered data. The 13th value in the ordered data lies in the group $20 < t \leqslant 25$. The modal group is the group with the highest frequency. The group with the highest frequency is $20 < t \leqslant 25$. This is the same group as the group that contains the median. So Franz is right.

Histograms (page 93)

1 Accurate histogram.

2 a 40   **b** 649.6 cm (649-650 cm)

Mixed exam-style questions (page 95)

1 a Accurate scatter diagram.

b (70, 70). Does not fit the pattern of the other data.

c Positive correlation. The longer the time taken to do the 250 piece jigsaw the longer the time taken to do the 500 piece jigsaw.

d i 82 – 84 minutes

   ii Reliable for data as interpolation, but may not be reliable overall as small sample of students.

2 a 22 000   **b** 15 000   **c** 25%

3 a 0.22   **b** 77 (77.6)

4 Accurate histogram.

5 $\frac{11}{23}$ **6** $\frac{33}{82}$

Working with stratified sample techniques and defining a random sample (page 97)

1 Brewyn 3 members, Dafddu 7 members and Cae Ben 6 members

The multiplication rule (page 99)

1 a Accurate tree diagram.   **b** 0.005   **c** 0.045

2 $\frac{11}{42}$

The addition rule and Venn diagram notation (page 100)

1 a $\frac{6}{23}$ **b** $\frac{16}{23}$ **c** 0

2 a Accurate Venn diagram.

   b $\frac{22}{67}$ **c** $\frac{49}{67}$

3 a 0.4   **b** 0.9

Conditional probability (page 102)

1 a i $\frac{8}{23}$ **ii** $\frac{3}{23}$ **b** $\frac{3}{8}$

2 $\frac{3}{7}$ **3** $\frac{42}{54}$

Mixed exam-style questions (page 103)

1 3064   **2 a** $\frac{1}{30}$ **b** $\frac{1}{3}$

3 $\frac{64}{105}$ **4** 0.032

5 a Accurate Venn diagram.   **b** 3

   c i $\frac{3}{24}$ **ii** $\frac{15}{24}$